CONNIE CAO

Your

# ASIAN VEGGIE PATCH

A guide to growing and cooking delicious Asian vegetables, herbs and fruits

# Hello!

Welcome to the start of your Asian veggie patch journey! I love to share the joy of growing Asian veggies at home, and to motivate as many people as possible to embrace and celebrate the diversity of edible plants grown and eaten across Asian households, beyond what you see in local supermarkets. So, in this book you'll find all the information you need to start growing, harvesting and cooking some of my favourite Asian veggies, herbs and fruits, no matter how big or small your garden space is.

This book is inspired by my Asian heritage, and it offers companionship and gentle guidance on your journey. Let me be your Asian best friend who can't stop raving about how celtuce is the next big thing, fresh jujubes are delicious, and sweet potato leaves should be eaten not chucked out. Expect bountiful encouragement and enthusiastic cheering from me on the sidelines.

# My story

I'm Connie, and I was born on Wurundjeri Country in Melbourne, Australia, to Chinese immigrant parents.

My parents grew up in Shanghai and came to Australia in 1988 in search of a better life. As new immigrants, they were isolated from their Chinese culture. They sought comfort in any Chinese food that would remind them of home.

My mum is a fabulous cook. When I was a child, she regularly made Asian dishes – although she'd often have to travel many suburbs away to buy the fresh produce she was familiar with from her life in China. Most of the time, my sister and I had no idea what we were eating. All we knew was that it was delicious.

My dad loves to garden and is always pottering about outside. When I was in my teens, my family moved to a house with a big backyard. My mum joined a Chinese social club, and it was there that she received her first packet of bok choy seeds from a new friend. My dad then planted his first Asian veggie. They were thrilled! It was a taste of home in our own backyard, and the beginning of a thriving family Asian veggie patch.

I watched my dad cultivate his patch, and saw how chuffed he was to grow veggies from his childhood and culture. I'd always dreamed of growing my own food, so I gardened alongside him. Both of us learned from scratch; I searched the internet for advice, and he used trial and error as well as pragmatic logic. In the kitchen, my mum would turn our harvests into easy-to-make Asian dishes.

The thing that got me into gardening is the magic that happens from seed to food. I still remember when I fell in love with plants: it was in primary school, when we planted chive seeds in a small pot. Before I knew it, I had a beautiful plant to look after. I was in awe then, and I'm in awe now. Nature reminds us that even the smallest things can be incredibly resilient and grow into something wonderful.

In a bit of a detour from my original law and commerce degrees, my unwavering passion for nature inspired me to achieve the Diploma of Sustainable Living as well as the internationally recognised Permaculture Design Certificate (PDC). Both courses gave me technical knowledge, plus a deep appreciation for what we've been gifted on this planet. This guides my organic gardening philosophy today.

If you're new to gardening, just know that organic gardening might come with imperfect(ly perfect) produce, and that's completely okay. I love imperfect veggies. They should be embraced because they are perfect just the way they are.

# Learning and growing

The best part of a home is always the veggie patch. The feeling of being outdoors and growing your own food is so special. When my husband, Tom, and I moved into our first home, I couldn't wait to plant my own veggie patch. I started off with whatever was most accessible, but I quickly came to miss growing Asian veggies. So, I asked my dad for some seeds. Now, among the flowers, berries, Mediterranean food plants and fruit trees, I grow delicious veggies from my heritage and create simple, home-style dishes with them in the kitchen.

Asian veggies are just as easy to grow as other veggies. My favourites are purple bok choy, mini wombok and winter melon, some of which have grown to 7 kilograms (over 15 pounds). I love to experiment with subtropical veggies as well, by making the most of warmer spots in my garden and my backyard greenhouse. It brings me so much happiness to reconnect with my culture through gardening.

Over the years, my garden has been my greatest teacher. I have learned not just from writing down my gardening observations, successes and failures, but also from simply gardening. My garden teaches me gratitude, patience, appreciation of the little things, self-compassion, resilience, joyful living and the beauty of imperfection. It's incredible that these teachings can be applied so meaningfully to daily life, helping me to cultivate a nurturing outlook on the world around me.

Gardening has also instilled in me the art of savouring the moment – embracing the tranquillity in the buzzing bees and sweetly fragrant air on an early spring day, no matter how busy life feels. But most of all, gardening is a pleasure! And a small reminder of the interconnectedness of everything in this world.

Growing an edible garden is a continual learning journey, and I love the idea that no gardener will ever know everything. Don't be afraid to experiment and make mistakes here and there – it's all part of the experience. You learn the most from your errors.

The most exciting thing about my gardening journey has been observing and analysing my plants. It's interesting to take a close look at plants, and I have made extensive notes throughout the seasons via words, pictures and videos. In this book, I use my gardening experience and my family's stories and favourite recipes to show you how to create and enjoy a thriving Asian veggie patch.

I hope this book helps you to discover the wonderful world of Asian veggies, herbs and fruits, and inspires you to grow your own Asian food plants at home. I'd love to hear about *your* Asian veggie patch! Please send me photos – I can't wait to follow your journey.

With love and sunshine,

# INTRODUCTION

There are so many wonderful reasons to start your own veggie patch, from reducing food miles and wastage, to eating seasonal, fresh and pesticide-free veggies. And then there's the incredible, life-giving intangibles: the sheer awe of being immersed in nature and nurturing your physical and mental wellbeing. It's also a practice of gratitude. Connecting closely with your food sources and the growing process is a great reminder that nature has to get a lot right to produce the food we often take for granted. But there are even more reasons to start your own Asian veggie patch!

**You can try unique and interesting varieties** There is a wide array of veggies, herbs and fruits out there, beyond what is available in your local supermarket, and this is especially the case when it comes to Asian veggies.

**It helps you to embrace and celebrate cultural diversity** Whether you have an Asian heritage or not, growing Asian veggies is a wonderful way to immerse yourself in the world of Asian food and culture. It's also a lovely way to appreciate and honour the incredibly diverse and multicultural community of which we are so lucky to be a part.

## Asian veggies can be fabulous permaculture plants (see page 14)

Many Asian veggies are hardy, adaptable and easy to grow without much attention. A large portion of them are also perennials (see page 21) and can serve multiple functions in the garden.

## You have easy access to fresh veggies all year round

Asian veggies can be difficult to find in the supermarket and, when it comes to freshness, it's hard to compete with backyard-to-kitchen veggies. Grow your own to ensure that you have fresh Asian veggies whenever you want them. In addition, harvesting your own homegrown veggies means that you have so much more of the plant available to use. This opens up a world of possibilities, as you can cook the stalks and young shoots as well as the leaves.

## Asian veggies can be nutritious

Adding Asian veggies to your meals can boost your intake of essential nutrients. For example:

- **Celtuce** contains double the amount of potassium and magnesium found in other lettuces. Depending on the soil it's grown in, it can contain up to five times as much vitamin C as other lettuces.
- **Water spinach** is a great source of vitamin A (50 grams, or less than 2 ounces, is enough for your daily vitamin A intake), iron and magnesium.
- **Bok choy** is high in vitamins (A, B6, C and K) and minerals (calcium, folate, magnesium, potassium, manganese and iron).
- **Broad beans** have high levels of protein (more than most other beans), amino acids and iron, making them a great source of plant protein for vegetarians and vegans.

## Asian veggies, herbs and fruits can have therapeutic qualities

Some are used in Traditional Chinese Medicine (TCM) for their medicinal properties. For example:

- **Chrysanthemum** tea is said to reduce blood pressure, inflammation and anxiety.
- **Goji berries** are said to improve immune system function, circulation and eyesight.
- **Ginger** can be used to boost immunity and help with colds, motion sickness and nausea.

## Asian veggies are delicious

Every food plant has its own unique flavour profile, and including Asian veggies in your kitchen line-up allows you to incorporate an even larger range of flavours in your dishes. Each and every Asian veggie is oh-so-tasty in its own way!

# What you will find in this book

To help you start your Asian veggie patch, I've included a section on gardening essentials (see page 18) for all levels of experience. It contains everything you need to get growing - from climate and soil to seed raising and pest issues. If you have gardening questions while reading this book, flip back to this section for more information.

I have then curated more than 40 of my favourite Asian veggies, herbs and fruits that are worthy of a place in your garden. Each plant profile is divided into three main sections: growing, harvesting and cooking.

## GROWING

I'll let you in on the best way to get started, whether it's from seed, seedlings or cuttings. Expect easy-to-follow advice on how to be the best parent for your plant, all explained in a beginner-friendly way. I'll also share plenty of small-space and container growing ideas, especially for those who, like me, live in urban areas. You'll also find permaculture gardening tips throughout, and handy hacks on how to get the most out of your plants.

## HARVESTING

When you grow your own veggies, you can harvest them however you like. I'll share all the great harvesting options that are available when you grow your own plants, as well as extra information on the lesser-known parts of plants that are also edible.

## COOKING

The third section is all about cooking with your harvest. I'll introduce you to some of my favourite Asian home-style recipes and ideas to help you get the most out of the plants you've grown. Many of these recipes have been inspired by my Shanghainese/Chinese heritage and are easy to cook at home. Others are influenced by my extended family in Southeast Asia and by my travels.

You'll find that most recipes focus solely on the plant in question, as often we gardeners harvest a lot of the same thing at the same time. Complementary ingredients are chosen based on what else is in season. I'm not one for fancy recipes, so expect easy-to-make dishes for busy people and modern lives.

Most recipes are for dishes that are served communal-style with rice on the side. I suggest cooking two or three dishes to share between two people, and three or four dishes to share between four people.

Also, the measurements are based on using fresh ingredients (for example, fresh ginger and fresh turmeric, rather than dried/powdered versions). If you want to try a recipe but don't have fresh ingredients on hand, then simply use one teaspoon of the dried/powdered version in place of three teaspoons of the fresh ingredient. For fresh leafy herbs, use half the amount of dried herbs.

Head to pages 58–61 for a list of my favourite Asian pantry staples and kitchen essentials.

# A permaculture way of gardening

As well as learning lots about Asian veggies in this book, you'll also discover more about permaculture. This holistic approach guides my gardening style, as it's a more sustainable, resilient and regenerative way of living. In permaculture, we look to Mother Nature as our greatest teacher, seeking to tread a path gently alongside her, while striving to have our needs for food, energy and resources met. When it comes to gardens, it's about designing them in a way that allows humans and natural ecosystems to coexist and thrive in harmony.

A portmanteau of 'permanent' and 'culture', permaculture was developed in Australia in the 1970s by Bill Mollison and David Holmgren. Permaculture encourages us to see that if we want to remain permanently on this planet for centuries to come, then we must cultivate cultural attitudes that will allow this to happen.

To me, permaculture design just makes sense. Humans are lucky to be the stewards of this diverse and life-sustaining planet, so we must look after it – because we are nothing without the natural ecosystems that support us: the air we breathe, the plants we eat and the resources we consume. We simply cannot take these things for granted. It is up to us to actively choose to protect what we have.

There are three overarching ethics and twelve guiding principles in permaculture. In the following pages, I've included a brief introduction to these ideas and how they apply to the way we garden. As you read through this book, you'll find gardening tips inspired by these ethics and principles. I hope they encourage you to adopt permaculture practices in your garden.

## THE ETHICS OF PERMACULTURE

**EARTH CARE** reminds us of the importance of looking after our natural world, and to take part in ways that can rebuild and protect the environment. This ethic points to the value of growing food using organic practices, and protecting biodiversity.

**PEOPLE CARE** prompts us to nurture ourselves and the community, as healthy people and communities are more resilient. This ethic encourages us to garden for health and wellbeing, both for ourselves and for others.

**FAIR SHARE** inspires us to be conscious of the resources we use, and to look after and distribute surplus to provide for one another where we can. One of the simplest ways we can practise this ethic is to share our excess harvests with others.

# THE PRINCIPLES OF PERMACULTURE

These principles can be applied to homes, neighbourhoods and communities at large. Here, though, I have focused on the main gardening aspects of the permaculture principles.

## OBSERVE AND INTERACT

Take the time to slow down and connect with your garden space. Get a good feel for what happens in your garden throughout the year (for example, where sunlight falls and which animals visit), then use this information as a guide when you create your Asian veggie patch. I make suggestions related to this principle in the book, such as mapping out the sun's movement across your garden, and experimenting with different planting locations.

## CATCH AND STORE ENERGY

Take advantage of abundant times by ensuring that useful things (such as solar power, rainwater and generational knowledge) are captured and stored to sustain you through the ebbs and flows of life. In a garden context, you can make the most of seasonal gifts through the simple art of preserving food by freezing, dehydrating, fermenting or bottling.

## OBTAIN A YIELD

We're all here to grow some deliciously good Asian veggies. This entire book is devoted to tips and tricks for making the most of your garden and turning it into a productive patch.

## APPLY SELF-REGULATION AND ACCEPT FEEDBACK

The finite nature of valuable resources encourages us to live within our means. Reflect on this by keeping a garden diary. Take note of what you're consuming when gardening, and think about ways to use what you already have on hand. Also, keep track of what did and did not work in the garden, so you can apply what you learn to future seasons.

## USE AND VALUE RENEWABLE RESOURCES AND SERVICES

There are many renewable resources out there that we can lean into. In the garden, compost piles turn kitchen and garden scraps – even cardboard boxes – into incredible organic matter for the soil. When it rains, catch the rainwater (either in a special tank or in a few strategically placed buckets) and use it on the garden instead of letting it flow away.

## PRODUCE NO WASTE

There are so many resources in our homes and gardens that go to waste. Think circular (re-use) instead of linear (buy and discard) where you can. In this book, I'll share ideas on how to make the most of the things we have, such as upcycling or repurposing common household items for use around the garden.

## DESIGN FROM PATTERNS TO DETAILS

If we look at the big picture first, then we can be guided by seasonal patterns, climates and even life stages. Tune in to your gardening patterns, such as how often you harvest certain things. Let these guide you through the specifics of planning where and when to grow certain plants, in order to create a more user-friendly garden.

## INTEGRATE RATHER THAN SEGREGATE

In nature, all things are intertwined. Plants don't just exist on their own – they lean on each other for support. I love the idea of elements in the garden being connected to each other. You might plant a shade-loving herb under a fruit tree, which could also provide a cool spot for chickens in summer.

## USE SMALL AND SLOW SOLUTIONS

Treading gently is kinder to the environment, more sustainable and easier to maintain than swift solutions. The simple and unhurried practice of collecting your own seeds leads to stronger plants that are better adapted to your local climate.

## USE AND VALUE DIVERSITY

Natural ecosystems brim with diversity above and below the soil, and so should our gardens. Throughout this book, you'll find plenty of nods to companion plants, edible perennials and pollinator-friendly choices that will provide habitats for diverse wildlife in your garden.

## USE EDGES AND VALUE THE MARGINAL

As an overly enthusiastic urban gardener, I'm always thinking of ways to fit in as many plants as possible. You can use the often overlooked spaces around your home, such as narrow walkways, concrete porches and vertical surfaces; as a bonus, these spots usually have special characteristics, such as a warm microclimate.

## CREATIVELY USE AND RESPOND TO CHANGE

Change is the only constant in life, but we can adapt in outside-of-the-box ways. Embrace change by doing your best to make the most of any given situation. Sometimes you'll find that the solution is in the problem itself. For example, a boggy corner of your garden is the ideal place to grow water-loving plants.

*'Stacking functions' – ensuring that every garden element serves multiple purposes – is a valuable permaculture design principle, especially when trying to make the most of a small space. For example, shark fin melon (a warm-season climbing vine) can be grown over a pergola to create shade in summer while it provides delicious food to eat.*

# PART ONE
# Gardening
# ESSENTIALS

If you're just beginning your gardening journey, then this section is for you! I've included plenty of key information to get you started. More importantly, this section will always be here for you – refer back to it whenever you need clarification about plants and their needs.

Within this section, I dive into climate, plant types and families, the importance of good sunlight and healthy soil, how to feed and hydrate your plants, and ideas on small-space and container growing, as well as the nitty-gritty about growing, harvesting and cooking your tasty Asian veggies. There are lots of interesting tips and tricks here, no matter where you are on your gardening journey.

That said, if you've been gardening for years and want to skip over this section, then feel free to head straight to the plant profiles (see page 62) to start planning your Asian veggie patch.

# Climate

Knowing your climate is the starting point for growing an abundant garden. It helps you understand what plants can be grown in your area and when you can grow them. For the purposes of this book, I have divided climates into three broad groups, guided by their frost tolerance.

When I talk about **warm climates**, I'm referring to subtropical, tropical or hot, arid climates that do not experience frost, and that generally have warm weather throughout the year. Picture places such as Brisbane in Australia and Miami in the United States.

When I refer to **temperate climates**, I'm thinking of areas that experience a balance of hot summers and cool winters with mild frosts. The place where I live – the Australian city of Melbourne – has a temperate climate, as do Sydney, Adelaide, Perth and Tasmania, most of the United Kingdom, and cities such as San Francisco in the United States.

When I say **cool climates**, I mean anywhere in the world that experiences heavy frosts or short growing seasons – for example, New York in the United States.

However, it's important to know that climates around the globe can be quite varied, and even two gardens in the same city or state can have different climates. There are also ways of growing annual veggies that take into account your climate (see page 22). If you need more guidance about the plants that can be grown in your area, then check with Asian veggie gardeners or nurseries near you – most gardeners love nothing more than to share their knowledge.

## FROST TOLERANCE

Frost is the thin layer of ice that forms on outdoor surfaces whenever the ground temperature reaches 0 degrees Celsius (32 degrees Fahrenheit). Even when the weather forecast reports the temperature as a few degrees higher than this, frost can form. This is because temperatures taken for weather forecasts are measured by devices located above the ground, where the temperature is warmer, and not situated on the surface of the ground, where the temperature is cooler.

Frost tolerance refers to whether a plant will survive when exposed to frost. There are three levels of frost tolerance:

1. **Frost-hardy** plants can survive heavy frosts, but the degree to which they will varies. Some plants are only hardy to −10 degrees Celsius (14 degrees Fahrenheit), while others can cope with −30 degrees Celsius (−22 degrees Fahrenheit).

2. **Frost-tolerant (or semi-hardy/half-hardy)** plants will survive light frosts – when temperatures drop just below 0 degrees Celsius (32 degrees Fahrenheit) for only a few hours – but will die if hit by anything more.

3. **Frost-sensitive (or frost-tender)** plants cannot survive frosts.

Knowing the frost tolerance of plants as well as the likelihood and level of frost in your area will help you to determine what can be planted in your garden and when it can be planted.

While frost-hardy plants survive frost, they don't grow much during this time. So if you live in an area with long periods of heavy frosts (for example, northern parts of the United States), then you'll probably need to take a gardening break during the winter months and just focus on the warmer growing season. If you live in an area with mostly mild frosts or no frosts (such as most parts of mainland Australia), then you'll be able to grow Asian veggies throughout the year.

## Plant types

Plants can be divided into three types, based on their life cycle: annual, biennial or perennial. Knowing a plant's life cycle can help you manage your expectations. For example, the first time I grew bok choy, I was heartbroken to see it produce leafy greens for only a couple of months before flowering, instead of growing more leaves. If I had realised that this was its natural life cycle, then I would have felt much better.

**Annuals** are plants that live for less than one year. They will germinate, grow, flower, set seed then die all in one growing season. Some annuals, such as pumpkin (squash), have a life cycle of up to five months. Others, including choy sum, complete their entire life cycle within two months.

Many Asian veggies (and veggies in general) are annuals, so new seeds or seedlings need to be planted every year. However, in return for this effort, annuals grow quickly and are highly productive. This is because they have only one season to get everything done, so they've evolved to do as much as possible before their time is over.

**Biennials** are plants that live for two years. They spend their first year growing and their second year flowering and setting seed. Unless they want to save the seeds, most gardeners grow biennials – such as Chinese broccoli and bok choy – as annuals. Some biennials will still flower and set seed in their first year anyway, depending on their growing conditions (this is what happened with my first bok choy crop!).

**Perennials** are plants that live for three years or more. Some perennials (such as chokos) are short lived, with a lifespan of up to seven years; others (such as persimmon trees) can live for 75 years. Because perennials have a longer life cycle, they often grow slowly; for some, it may take a few years before they bear fruits – although perennials that bear fruits in their first year do exist (such as eggplants/aubergines). Perennials are gentler for soil ecosystems, as you're not continually disrupting underground life by digging the soil and planting new seedlings each year. This also means less work for you. All of these benefits make edible perennials well-loved in permaculture gardens.

*Short-lived annuals are great for filling in temporary gaps between young and growing perennials. Longer-living perennials are ideal for parts of the garden you visit less often, as they are less hands-on and don't need to be replanted each year.*

## MORE ABOUT ANNUALS

Annual veggies can be further divided into warm-season and cool-season plants.

### WARM-SEASON ANNUALS

These are the veggies you grow in summer. They are frost sensitive, prefer warmer temperatures and grow very poorly in cool weather. Examples of warm-season annuals include winter melon and loofah.

Warm-season annuals can be grown when daytime temperatures are consistently above 22 degrees Celsius (72 degrees Fahrenheit) and overnight temperatures are at least 10 degrees Celsius (50 degrees Fahrenheit). In warm climates, this could be the majority of the year or even all year round.

For temperate and cool climates, warm-season annuals are typically transplanted into the garden in spring, after the last frost, and will grow through summer. It's best to wait for a week's worth of minimum overnight temperatures of 10 degrees Celsius (50 degrees Fahrenheit) before planting out warm-season annuals.

As temperate- and cool-climate gardeners have a shorter period of warm-weather growing, there is less time for slower-growing warm-season annuals to reach maturity. If this is you, then there are two ways to make the most of your growing season:

1. **Start your seeds indoors** – Sow seeds and raise seedlings in your house or in a greenhouse (if you have one), where the temperature is warmer, before transplanting seedlings into the garden. This tricks your plants into thinking that summer has begun early. Seeds can be started this way one or two months before your last frost, then slowly acclimatised to the outdoors (see page 50) before planting in the garden when the temperature is warm enough.

2. **Protect your plants from cold and frost** – Once the weather starts to cool, warm-season annuals will slow down their growth. However, you can trick them into thinking that summer's still on. Provide extra warmth by moving them into a greenhouse or indoors, using a cold frame or growing your plant next to a brick wall that receives afternoon sunlight. These simple techniques can also be used to keep plants safe during periods of unexpected cold weather at the start of the season.

### COOL-SEASON ANNUALS

These are veggies that prefer to grow in cool temperatures. They are tolerant of frosts, but some tolerate light frosts only; others are very frost hardy. Some cool-season annuals even taste sweeter when exposed to frost. Once warm weather hits, many cool-season annuals bolt – prematurely flower and set seed – which signals the end of their life cycle. Examples of cool-season annuals include celtuce and chrysanthemum greens.

Cool-season annuals prefer daytime temperatures around 20 degrees Celsius (68 degrees Fahrenheit). For warm climates, cool-season annuals are best grown through winter and located in a shady spot. For temperate and cool climates, cool-season annuals are usually sown at the start of autumn or at the start of spring.

*When buying warm-season annual seeds for cool-climate gardens, keep an eye out for varieties that are 'early maturing' or 'cold tolerant'. These plants either grow more quickly (ideal for shorter growing seasons) or cope better in cool climates.*

## BOLTING AND HOW TO PREVENT IT

Bolting is a panic response from plants; they're trying to reproduce quickly so that future generations of plants can live on. It's often triggered by hot weather, but it can also occur as a result of:

- fluctuating temperatures in general
- water stress or any stressors in general
- changes in day length.

The process of flowering and setting seed is natural, but it can be frustrating when it happens earlier than expected. It may shorten your harvest window or – even worse – occur before you get the chance to harvest anything at all.

Once plants bolt, and they start to focus their energy on producing flowers, their leaves become bitter and are no longer nice to eat. Even root veggies (such as daikon) can bolt, causing the roots to become bitter and woody. The best bolter I know is coriander (cilantro) – it's *always* on a mission to bolt.

If you struggle with bolting plants, then here are six tips to help you postpone the process for as long as possible:

1. **Select the right variety for your climate** – Different varieties have different temperature and climate preferences. If you're concerned about a certain species bolting in your climate, then look out for heat-tolerant or slow-bolt varieties.

2. **Experiment with planting times** – If you find that certain plants bolt quickly, then try planting them a month early. This can help you get a better harvest before the weather warms up. Or try both autumn and spring planting times, and see what works better for you.

3. **Try succession planting** – Sow a new batch of seeds or plant seedlings every few weeks. This ensures that you have a continual supply of young veggies in your garden, and older plants can be left to bolt.

4. **Reduce plant stressors** – Keep your plants consistently watered, especially during periods of warm weather. Ensure that your seedlings don't become root-bound before planting out. Always harden off seedlings before moving them to a new environment (see page 50 for information on hardening off).

5. **Plant in the shade** – If you live in a warm climate, it's best to plant cool-season annuals in shady spots. These areas have cooler temperatures than sunny spots, making them more bearable for cool-season annuals.

6. **Snip off any signs of bolting** – As soon as you see any suspicious-looking bolting behaviour (such as a flower bud in the middle of your plant), you can attempt to slow the process by snipping off the growth. This can extend your harvest for a week or so, but it won't stop your plants from bolting altogether. I find that once plants have entered bolt mode, they're determined to follow through … no matter what.

## EMBRACE YOUR CLIMATE!

Sometimes, we've got to embrace nature just as it is instead of fighting against it. Many plants bolt as the weather warms up, at a time when there are minimal flowers in the garden. Letting your plants flower at this time of the year helps to attract pollinators and provide valuable food for them; in turn, the pollinators help you by pollinating your veggies.

If you live in a warm climate where plants bolt before you can harvest them, then it's helpful to remember that every single plant has its own individual preferences (just like humans!), and not everything can be grown everywhere. Sometimes, it's best to accept that some plants aren't suited to your climate. Instead, focus on curating a list of plants that do thrive in your climate. Growing these plants in abundance is a better use of your space and energy than trying to coax a plant out of its comfort zone.

## MORE ABOUT PERENNIALS

Perennial plants can be divided into groups based on their frost tolerance.

### FROST-HARDY AND FROST-TOLERANT PERENNIALS

These plants will either hang on to their leaves all year round (evergreen) or shed their leaves during the colder months (deciduous), pushing out fresh new growth as the weather warms up in spring. For some plants (such as jujubes and nashi pears), a certain number of chill hours – cold weather under 7 degrees Celsius (45 degrees Fahrenheit) – is essential in order for them to come out of winter dormancy, flower and set fruit.

### FROST-SENSITIVE PERENNIALS

These include lemongrass, eggplant (aubergine) and Malabar spinach. Most frost-sensitive perennials are native to warm, subtropical or tropical climate areas, such as Southeast Asia, India or southern China. They're usually evergreen in warmer climates, but they can die down in winter when grown in cooler climates and re-emerge as the weather warms up.

While frost-sensitive perennials grow easily in warm climates, gardeners in temperate and cool climates can also grow these plants using the strategies below.

1. **Grow them as annuals** – Fast-growing frost-sensitive perennials (such as eggplant/aubergine, chilli and Malabar spinach) can simply be grown as annuals, and can be harvested before the first frost hits. If you live in a temperate or cool climate, then you might not even realise that some of the plants you grow as annuals are actually perennials, since this practice is so common.

2. **Protect them in winter** – Grow your frost-sensitive perennials outdoors in summer, move them to a protected spot in winter, then pop them outside again the next summer to allow them to continue growing. In winter, you can move your plants indoors, into a greenhouse, to a sheltered area such as a carport, under the eaves of a house or against a sunny brick wall where the temperature is warmer.

3. **Treat them as biennial-bearing plants** – Some plants (such as ginger, turmeric and galangal) need a long, warm growing season to have anything worth harvesting. These plants are a little more challenging to grow if you live in a temperate or cool climate, but you can allow them to grow over two summers by keeping them protected in winter and only harvesting them every second year.

# Plant families

Every plant belongs to a family and shares characteristics with other plants in the same family. For example, chrysanthemum greens and celtuce belong to the daisy family (Asteraceae), Thai basil and shiso are part of the mint family (Lamiaceae), and sweet potato and water spinach can be found in the morning glory family (Convolvulaceae). Knowing to which family a plant belongs can usually give you a helpful hint about a plant's characteristics. Below, I've listed some of the more popular families that include Asian veggies, and their common traits.

## CUCURBIT FAMILY (CUCURBITACEAE)
Members include winter melon, hairy melon, loofah and pumpkin (squash). Here are their characteristics:
- frost-sensitive, warm-season plants that are grown over summer
- large plants that mostly grow as vines, but some grow as bushes
- male and female flowers on the same plant that require pollen transfer from male flowers to female flowers in order to produce fruits (done by pollinating insects, or aided by hand-pollination; see page 153)
- heavy feeders and heavy drinkers that require lots of organic matter and water
- large, broad, edible leaves and large, (usually) yellow, edible flowers
- large, mostly cream-coloured seeds shaped like a pepita (pumpkin seed)
- can be susceptible to powdery mildew towards the end of the season.

## LEGUME FAMILY (FABACEAE)
Members include broad bean, snow pea (mange tout), long bean and flat bean. Here are their characteristics:
- family contains both warm-season plants (such as flat bean and long bean) and cool-season plants (such as broad bean and snow pea/mange tout)
- many grow as vines and need vertical support, but some grow as bushes
- self-pollinating flowers become legumes (such as beans and peas)
- light feeders but appreciate water
- seeds are mature dried beans or peas.

## MUSTARD FAMILY (BRASSICACEAE)
Members include bok choy, wombok, daikon and Chinese broccoli. Here are their characteristics:
- cool-season plants, with varying levels of frost tolerance/hardiness
- sensitive to temperature and environmental stress, which can cause plants to bolt
- whole plant is edible, including leaves, stems and flowers
- mostly medium to heavy feeders, wanting lots of nitrogen
- fast growers that appreciate being well-watered
- small, round, brown seeds (they look like the mustard seeds used as a spice in cooking)
- can be susceptible to cabbage moth and cabbage butterfly caterpillars, slugs, snails and aphids, as well as white leaf spot (a fungal disease).

## NIGHTSHADE FAMILY (SOLANACEAE)

Members include eggplant (aubergine), chilli and goji berry.
Here are their characteristics:

- frost-sensitive, warm-season plants that are grown over summer
- each flower has both male and female parts, and is self-pollinating
- heavy feeders and heavy drinkers
- grown for their fruits; leaves are usually inedible and can be toxic
- mostly small, cream-coloured seeds (they look like chilli seeds).

## ONION FAMILY (ALLIACEAE)

Members include garlic, spring onion (scallion) and garlic chives.
Here are their characteristics:

- frost-hardy plants that are usually grown during cooler weather
- small plants with long, grassy, vertical stems and fragrant round flower heads containing lots of little flowers
- mostly medium feeders and medium drinkers
- plants are grown for either their aromatic underground bulbs (such as garlic) or their fragrant leaves (such as spring onion/scallion and garlic chives)
- small, black seeds
- can be susceptible to black aphids (small, sap-sucking insects that aren't particularly welcome in edible gardens!).

### NITROGEN FIXING

Members of the legume family are renowned nitrogen fixers and can assist in nourishing the soil with nitrogen (the most important nutrient plants need to thrive). Aided by symbiotic bacteria, legumes have the ability to capture nitrogen from the atmosphere and store it in their roots. (Legumes don't need much fertiliser because they feed themselves with nitrogen!) At the end of the growing season, when the plants are cut back to the ground, they decompose and release the remaining nitrogen into the soil. Because of this, legumes make a great green-manure crop to feed next season's plants.

# Sunlight

All plants need some level of sunlight to thrive. Some require full sun, while others cope with a part-sun position. A small number tolerate shade (as long as they receive indirect light or a tiny bit of sunlight at some point during the day).

**Full-sun** plants require direct sunlight hitting their leaves for at least six hours a day. A north-facing spot in the southern hemisphere (or a south-facing spot in the northern hemisphere) is a great example of a full-sun location. In urban areas, keep an eye out for large trees or buildings that may cast shadows on your plants during the day. Most fruiting veggies (such as pumpkin/squash, shark fin melon and loofah) prefer a full-sun position.

**Part-sun** plants (also known as part-shade plants) require four to six hours of sunlight a day. They prefer morning, east-facing sunlight, as it's cooler; afternoon, west-facing sunlight can be too warm and intense for these veggies. Many leafy greens (such as bok choy and Asian mustards) do well in part-sun.

**Shade-loving** plants are less common, but they do exist – wasabi is one. Its natural habitat is shady riverbanks in the mountain valleys of Japan. Shade-loving plants usually live in the bottom layer of forests, enjoying dappled or indirect sunlight during the day. In an urban garden, they can be planted beneath the canopy of fruit trees or within the shade line of a fence or building.

# HOW MUCH SUNLIGHT DO YOU HAVE?

A good way to figure out how much sun a certain spot in your garden receives is to take a photo of it every two hours on a sunny day. Do this in both summer and winter, as the sun's angle changes throughout the seasons. Spots that are sunny in summer but shady in winter are great for deciduous trees, as they are dormant in winter, or can be used for summer plantings and then left uncultivated during winter.

Another way is to simply experiment. I do this by putting my plants in pots first, then placing them in the desired location and seeing how they go. If they're happy there, then I'll plant them in the ground. If not, then I'll try somewhere else.

If you're reading this and thinking, *I don't think my garden has enough sun at all* – don't worry, it's going to be okay. You can still grow a wonderful veggie patch! While full sunlight can be difficult to come by if you live in a built-up urban area like I do, you can still make the most of the space you have.

My garden doesn't receive full sunlight throughout the seasons (although luckily for me, I do live in Australia, where it's sunnier than most places), so I use the following guidelines to help me grow an abundance of food.

## FOCUS ON PLANTS THAT SUIT THE SPACE

We'll get the best results when we tailor the things we grow to the conditions we have in our garden. Every garden can be celebrated for its special growing environment and the unique mix of veggies it can grow. For example, having a shady garden can be a good thing. It means that there's a chance you can grow cool-season leafy greens and herbs during the warmer months of the year, when most other gardeners will struggle because of too much sun. You can then swap veggies with another gardener, so you both have a variety of veggies to enjoy.

## ONLY GROW DURING SUMMER

During summer, most gardens have better access to sunlight because the sun is high in the sky. Focus on having a bumper growing season during this time, then take a rewarding break during winter. You can make the most of this quiet time by getting in some deep rest, planning out your next season or doing some indoor growing (see page 55 for some ideas).

## GIVE IT A GO ANYWAY

If the space is going to sit empty, why not just try a plant there anyway? Many plants will still grow with less sunlight than they prefer – you'll just find that they grow a little less prolifically or a little more slowly than usual. This is what I do in my backyard during winter, when parts of my garden are covered by shade. I plant things anyway and keep my expectations reasonable.

# Soil

After you work out your sunlight levels, then the next thing to look into is your soil. When creating any kind of veggie patch, it's important to start with good soil. Most edible plants prefer fertile (nutrient-rich), well-draining soil.

## GARDENING IN THE GROUND

If your veggie patch is located directly in the ground, and it's your first time using that garden space to grow plants, then an excellent thing you can do for your soil (and future plants) is to add organic matter to increase soil fertility and improve soil texture.

**Organic matter** refers to any material made from plants or animals. This includes:

- compost
- aged animal manures
- mushroom compost (a by-product of mushroom farming)
- leaf mould (decomposed leaves)
- worm castings
- green manures.

*Green manures are specific types of plants that are sown densely, grown until just before they flower, then cut down and dug into the soil. This adds organic matter and nutrients to the soil, and is a great way to prevent weeds from growing in otherwise empty garden beds. See pages 29, 68 and 81 for examples of green manures.*

## CAN YOU USE ALL ANIMAL MANURES ON GARDENS?

**Chicken and cow manures** are the most common animal manures used in gardening. **Sheep manure** is less readily available but can also be used. All of these can be added directly to your soil – just let the manure sit for a couple of weeks before planting into it so the richness doesn't burn the delicate roots of baby plants.

**Pet poo** (from dogs, cats and other omnivores) should be avoided in edible gardens, as it can contain harmful bacteria and pathogens. It can, however, be composted first through a dedicated worm farm and then added to ornamental gardens. On the other hand, **rabbit manure** is fabulous, and you don't need to age it before using it in the garden. And because rabbits are herbivores, their manure is safe to use anywhere in the garden.

While **horse manure** is readily available roadside for a dollar where I live, a horse's digestive system does not break down weed seeds, so it's best to put the manure through a hot-compost system before adding it to the garden. Using it directly can introduce a bed full of weeds.

Adding organic matter is the magic key to productive and fertile soil for three main reasons:

1. Organic matter adds important and essential nutrients to your soil, such as nitrogen, phosphorus and potassium. These nutrients are slowly released as the organic matter decomposes over time, with the help of worms and microorganisms in the soil.

2. Organic matter improves soil structure, drainage and water retention. For clay soil, organic matter helps to break down clay to allow water to drain through better. For sandy soils, organic matter helps to bind sand particles, improving water retention. Organic matter also binds soil particles together in chunks so they are the perfect texture for holding nutrients, air and water.

3. Organic matter turns lifeless soil into a thriving ecosystem that is full of life. It contains fungi, insects, worms, bacteria and microorganisms, and it provides nutrients for these soil helpers, who in turn help to break down organic matter, suppress diseases and support biodiversity.

## GARDENING IN RAISED BEDS

Ensure that your garden has healthy soil from the get-go by creating your own nutrient-rich growing medium from scratch, rather than trying to make existing soil better. A fabulous way to fill a raised bed is to use the no-dig (lasagne) gardening approach, a permaculture way of gardening. Simply fill your raised bed with alternating layers of carbon-rich and nitrogen-rich organic matter. This will decompose over time to form a nutrient-rich growing medium, just like a decomposing forest floor. Follow these steps to fill your raised bed:

1. Start by placing a thick layer of newspaper or cardboard as the bottom layer. This helps to stop existing weeds or grass from growing, and it will naturally break down over time to provide carbon matter for the soil. (For raised beds on concrete, place some sticks, rocks or branches as your first layer instead, to help with drainage.)

2. Add 5–10 centimetre (2–4 inch) thick layers of carbon-rich inputs and nitrogen-rich inputs (see table opposite), alternating between the two until the bed is filled. Water in each layer as you go.

3. Add a thick layer of straw mulch on top to finish off the bed.

4. Dig holes slightly bigger than your seedlings, sprinkle compost into the holes, then plant your seedlings. This helps to protect your seedlings from touching any fresh manure that might burn their tender roots. Or wait for your bed to settle and decompose for a few weeks before planting.

Compost ↓      ↓ Compost

Straw mulch
Nitrogen
Carbon
Nitrogen
Carbon
Newspaper/cardboard

| CARBON-RICH INPUTS | NITROGEN-RICH INPUTS |
|---|---|
| • dried autumn leaves<br>• dried grass clippings<br>• egg cartons<br>• lucerne mulch<br>• pea-straw mulch<br>• sawdust<br>• shredded newspaper<br>• shredded plain cardboard<br>• sugar-cane mulch<br>• woodchips | • aged animal manure<br>• bokashi waste (see page 41)<br>• coffee grounds<br>• compost<br>• food scraps (best used in the bottom layers only)<br>• fresh grass clippings<br>• seaweed<br>• tea leaves<br>• worm castings |

## GARDENING IN CONTAINERS AND POTS

When growing veggies in pots, you should use a quality **potting mix** that has been specially formulated for container growing. It will often include a mix of:

- structural particles (such as aged woodchips)
- water-retaining particles (such as coconut coir)
- drainage-assisting particles (such as perlite and vermiculite)
- food for your plants (such as slow-release fertiliser and organic matter).

See pages 38–9 for more tips on growing veggies in containers.

## SOIL PH

Measured on a scale from 0 to 14, soil pH refers to how acidic or alkaline your soil is. Neutral soil is 7; anything below is acidic, and anything above is alkaline.

It's handy to know your soil pH because it affects plant nutrition. If a plant's preferred soil pH is different from the actual soil pH, then the plant's roots won't be able to access the nutrients they need from the soil – even if the soil is full of nutrients. Most veggies grow well in neutral soil, but some prefer a slightly acidic or alkaline soil.

## HOW TO CHECK SOIL PH

You can get reasonably priced kits for testing soil pH from your local nursery or garden centre – simply follow the provided instructions to gather a soil sample and add the dye indicator and powder, then reference the provided pH chart. Each test only takes a few minutes. A single kit can be used hundreds of times, so you can share it with your gardening friends.

If your soil is too alkaline and you want to make it more acidic, then add sulphur granules to your soil according to the packet directions. Note that sulphur works its magic slowly, so it might take a few months to change the soil pH.

If your soil is too acidic and you want to make it more alkaline, then add garden lime or dolomite to your soil according to the packet directions.

*One way to maintain the no-dig ethos and avoid disturbing the soil ecosystem is to simply cut your plants back to the base at the end of their life cycle (instead of pulling them out). Leave the roots to decompose in the ground, and they'll feed the wonderful living organisms in your soil.*

*Regardless of whether you're cultivating Asian veggies in the ground, in raised beds or in containers, don't forget to replenish your soil or potting mix with fresh organic matter each growing season.*

**SOIL PH SCALE**

- Alkaline (14)
- Neutral (7)
- Acid (1)

# Container growing

As someone who lives in an urban area, I'm always thinking about maximising my garden space. Growing in containers is a fabulous way to increase your growing space to include concrete patios, balconies, fences, walls and windowsills. Containers are especially great if you live in an apartment or don't have a garden, and they are easy to shift around to chase the sun or to bring to your next place when you move.

While growing in containers offers lots of flexibility, container plants do require extra love. They have no access to soil in the ground, so they depend on you to give them what they need.

## WATER

Container-grown plants need to be watered more often than in-ground or garden-bed plants. This is because containers hold less soil (and hence less water) and dry out more quickly on hot days.

When watering, make sure that you water the entire volume of soil until the liquid starts draining from the bottom. During very hot periods, small pots may need to be watered two or three times a day, especially if you live in a dry climate or it's windy. An easy way to ensure that container-grown plants get enough water on hot days is to use self-watering pots. You can also place saucers underneath your pots and fill them with water to provide your plants with extra liquid to soak up.

## FOOD

Container-grown plants only have access to whatever's in the potting mix, and they can't send out roots to forage for nutrients beyond that. So, they depend on us to supply the nutrients they need. I prefer to use a combination of slow-release fertiliser and liquid fertiliser:

- Apply an organic slow-release fertiliser (such as compost, well-aged animal manure or chicken manure pellets) to the potting mix at the start of the growing season. This ensures that the plants always have a basic level of food available to them.
- Give a dose of liquid fertiliser when watering every now and then. This provides the plants with an extra boost of instantly available nutrients.

## CHOOSING THE CONTAINER SIZE

Larger pots are easier to look after than smaller ones. They dry out more slowly, and they allow plants more access to nutrients and more room to grow a strong root system. When you think about it, large containers are pretty much the same as small garden beds!

A good way to manage your container growing is to cultivate a few small plants together in one large pot. For example, plant five spring onions (scallions) in a large container, rather than each plant on its own in a small pot. You can also group together different plants with the same needs, such as Thai basil and shiso.

## WHEN TO GROW IN A CONTAINER

Many Asian veggies, herbs and fruits lend themselves to container growing. Even if I had the in-ground space to plant them, I'd still grow them in pots. These include:

- **plants that need to be dug up to harvest** (such as ginger, turmeric and galangal) – simply empty the pot to gather the rhizomes
- **frost-sensitive perennials** (such as eggplant/aubergine and lemongrass) – you can easily move potted plants to a sheltered spot in winter
- **invasive and weedy plants** (such as jujube and goji berry) – they send out root suckers around the garden, so growing in a pot helps to contain the roots
- **super-producing plants** (such as curry leaf tree and yuzu) – these trees can grow large and produce more than you need when in the ground but will happily live in pots; this constrains their harvest to a more manageable amount for one household
- **plants that happily live in pots** (such as chilli and cumquat) – by growing them in pots you save your precious in-ground space for something else.

# Food and nutrients

Each plant draws a different amount of nutrients from the soil. Throughout this book, you'll see plants categorised as light, medium or heavy feeders, depending on how much nutrition they need to grow.

**Light feeders** include plants in the legume family, which have nitrogen-fixing abilities (see page 29). Some herbs and other small leafy greens also fall into this category.

**Medium feeders** include those in the onion family (such as garlic chives and garlic), daisy family (such as chrysanthemum greens) and a number of leafy greens and root veggies.

**Heavy feeders** include most fruiting plants, such as those in the nightshade and cucurbit families. Fast-growing crops (such as those in the mustard family) are also heavy feeders, as are any plants that produce lots of leaves in a short amount of time (such as Malabar spinach).

## HOW TO FEED PLANTS

The best way to ensure that your soil has adequate nutrients for your plants is to start each growing season by topping up your garden beds with fresh organic matter before planting. Use things such as compost, well-aged manure and worm castings.

One feed is usually enough to get light and medium feeders through the growing season. Heavy feeders will need a top-up or two as they grow. I usually feed heavy feeders every four weeks, especially when they're starting to flower or fruit.

During the growing season, keep an eye on your plants to see how they're tracking – a garden diary is perfect for making notes. Plants will naturally feed from the soil as they grow, depleting the soil of nutrients over time. If you find that your plants are growing slowly or not producing many fruits, or the leaves are starting to turn yellow, then these could be signs that the soil is low in nutrients and needs additional fertiliser.

*Weeds can be a nuisance in the garden, but they can be turned into a good thing. Make **weed tea** by soaking pulled-out weeds in a big bucket of water - be warned that this tea comes with a smell.*

*Cover the bucket with a lid to keep the mozzies out, and let it sit for a few weeks out of direct sunlight until everything (including any pesky weed seeds) breaks down.*

*Then dilute the weed tea (one part tea to ten parts water), and use it as a liquid fertiliser in the garden.*

## WHAT TO FEED PLANTS

The three most important nutrients that plants need are nitrogen (N), phosphorus (P) and potassium (K). In addition, plants also need a range of other micronutrients and trace minerals.

- **Nitrogen** helps with leaf growth and creates the green colour of leaves.
- **Phosphorous** helps with photosynthesis, root growth, flower development and seed production.
- **Potassium** helps plants to grow strong and fast, and assists with flower and fruit production, water uptake and disease resistance.

**Homemade organic fertilisers** are best, as they make the most of existing resources. Do-it-yourself compost is usually more potent and nutrient-rich than store-bought versions. Worm castings, green manure, diluted liquid from bokashi bins and weed tea (a great source of nitrogen) are all incredibly useful for adding nutrients to your soil.

With homemade fertilisers, you need to have enough ingredients on hand to make them yourself. Also, they're best used as more of an all-round fertiliser rather than one that targets a specific nutrient deficiency. So, sometimes you may need to supplement with store-bought fertilisers.

**Store-bought organic fertilisers** are another option. The good thing about them is that they're labelled with their NPK ratio, showing how much nitrogen, phosphorus and potassium the product contains. This allows you to better understand what you're adding to your soil. Organic fertilisers that you can find in garden centres or nurseries include bloodmeal, bonemeal, chicken manure pellets and fish emulsion. Store-bought compost, animal manure and mushroom compost are also available.

### MEET THE BOKASHI BIN

This is a Japanese composting method where you fill a small, bucket-like bin with layers of food scraps and a special grain, which anaerobically ferments the waste to give you a nutrient-rich liquid fertiliser that you can dilute and use. It also creates semi-decomposed organic matter that can be further processed in the garden or in a composting system.

One of the best things about bokashi bins is that almost any kind of food waste can be put in them, including meat scraps and small bones. The magic happens within an airtight container, meaning no smells – making it a fabulous option for indoor composting, small spaces and apartments.

# Water

Some plants need to be kept consistently moist, while others require regular watering but are happy to dry out for a day or two in between. Some plants are happy to sit in water, while others' roots will rot. For every plant in this book, I have flagged its individual water needs so you know how to best care for it.

## WHEN TO WATER

I personally prefer to water in the morning. In summer, this gives the water a chance to soak deep into the soil, reducing the chance of moisture loss through evaporation when the day heats up.

When watering, always water the soil and not the foliage. This, alongside morning watering, can help to reduce the risk of moisture-related fungal diseases that thrive in humid and damp environments. Also, water deeply to allow the liquid to seep lower into the soil. This encourages plant roots to travel downwards to access water, resulting in stronger and more resilient root systems.

## MULCHING

Mulching is a fabulously simple way to reduce the speed of water evaporation from the soil's surface. It serves other functions as well, such as insulating the soil from the summer heat and winter cold, and discouraging unwanted weed growth by preventing weed seeds in the soil from accessing sunlight to grow.

Organic straw mulches (such as pea straw, sugar cane or lucerne) are great choices for veggie patches because they naturally break down over time and enrich the soil with organic matter. Top up these mulches once or twice a year, at the start of each growing season.

## LIVING MULCHES

Another way to mulch is to use living plants. They can be used in conjunction with straw mulches to further reduce moisture loss and suppress weed growth. Some handy living mulches include:

- **sweet potato** – the edible leaves can be left to sprawl across the soil surface
- **alyssum** – a small, edible, flowering ground cover that attracts beneficial insects to the garden
- **nasturtium** – an edible flower that can be grown as a large, spreading ground cover; it can also be used as a trap crop to keep pests away from leafy greens.

*The best place to store water is in your soil. Adding organic matter to your soil can help to improve your soil's water retention and the amount of water that your garden will hold. Constructing swales (strategically located trenches) can help to capture and keep rainwater in your garden.*

*Rainwater is a precious resource that falls free from the sky, only to be washed away through stormwater drains. A rainwater tank is a wonderful investment and will help to capture this water for use in your garden during drier times of the year. Alternatively, you can collect rainwater in tubs, buckets or a wheelie bin.*

# Growing

Now you know the basics of climate, plants, sunlight, soil, water, food and nutrients, it's time to grow some Asian veggies! There are a few different ways you can start. These include buying seedlings, growing from fresh produce, sowing seeds indoors and directly sowing seeds in the soil. Each method has its own pros and cons.

## GARDENING STARTER KIT

Here are a few tools that you'll find handy for your Asian veggie patch.

### HORI HORI KNIFE

Originating from Japan, a hori hori knife is a multipurpose tool shaped like a trowel, but with a knife edge on one side, a saw edge on the other, and depth measurements marked on the blade. You can dig, plant, cut and even do light pruning with it. If you buy only one gardening tool, then make it this one.

### SECATEURS

Another must-have is a good pair of secateurs (pruning shears/hand pruners). These look like heavy-duty scissors and are used in the garden for cutting plants. They'll cut anything up to 2 centimetres (¾ inch) in diameter, including thin branches. They can also be used to harvest fruits and veggies.

### LOPPERS

Loppers look like giant secateurs with long handles, and they're ideal for pruning branches up to 4 centimetres (1½ inches) thick, such as fruit tree branches. If you have only a couple of trees that need pruning once a year, then why not borrow loppers from a neighbour and offer some of your delicious harvest in return?

### SMALL GARDEN FORK

It's handy for specific tasks, such as weeding, lifting garlic bulbs and separating plant roots before planting or repotting.

### SHARPENING TOOL

Secateurs, loppers and other tools with a cutting blade should be regularly sharpened using one of these. This helps to extend their life span. Also, sharp blades require less effort to use and result in cleaner cuts that heal more quickly.

### HARVEST BASKET

Manifest that delicious produce! Thrift shops often have beautiful vintage baskets, so please check them out before buying anything new. Alternatively, use what you already have: a large re-usable container, a salad spinner or a paper grocery bag.

### GARDEN DIARY

I recommend documenting your Asian veggie patch journey, as you'll learn so much from simply reflecting on your experience. Make notes of what's working and what's not. All of this will become valuable information for next season's preparations. A simple notebook will do; digital notes work, too.

## BUYING SEEDLINGS

Although a punnet of seedlings can cost more than a packet of seeds, if you're new to gardening, then starting with seedlings is a fabulous, stress-free way to begin your gardening journey. The seed germination has been done for you, and you're working with a more established plant. Once you feel confident in guiding seedlings to harvest, then you can challenge yourself by starting your plants from seed.

That said, buying seedlings isn't just for beginners. Seedlings are also a great option for cultivating plants with seeds that take forever to germinate, or if you don't have the time, space or supplies to raise seeds. They're also helpful for filling in unexpected empty spots that appear in your garden.

Nurseries usually only have the space to stock a limited variety of seedlings, guided by what's in high demand. Some of the more popular Asian veggies and herbs (such as bok choy, wombok and coriander/cilantro) are the most likely to be available as seedlings. Alternatively, check online marketplaces to see if local gardeners are selling home-raised seedlings.

## GROWING FROM FRESH PRODUCE

Surprisingly, you can propagate some plants from fresh produce that you've bought at the supermarket, and I'll offer some of these propagation hacks in the plant profile section. Starting plants this way is often easier than growing from seed, and these plants grow more quickly and can be harvested sooner. If you want to try any of the propagation methods I mention later in the book, ensure that you use organic, locally grown produce. And please check your local regulations to see if any restrictions apply.

## STARTING FROM SEED

Some Asian veggies aren't available as seedlings, so you'll have to buy seeds. Check the seed aisle of nurseries, or buy online from local seed suppliers. Join a gardening club or connect with local Asian veggie patch gardeners who can swap seeds with you. Sometimes, you can buy seeds (such as dried peas or broad beans) in bulk from the supermarket, which is cheaper than buying from a nursery. Again, it's best to choose organic and check if any local restrictions apply.

Seeds can be either directly sown in the garden or started in punnets, then transplanted into the garden when ready. Sowing direct is less work, but your plants may have a shorter growing season. Starting in punnets (usually indoors) has more moving parts (and a small upfront investment in equipment and supplies; see page 48) but can give your plants a beneficial head start in life.

## RAISING SEEDS IN PUNNETS

If you're new to starting seeds in punnets, you'll need:

- **seedling pots and trays** – collect and re-use the pots from nursery seedlings, and keep trays handy to use as water-drip trays
- **seed-raising mix** – a specially formulated, finely sieved soil mix for germinating seeds
- **watering-can with a gentle spray** – using a heavy flow can wash away small seeds
- **vermiculite** – a natural mineral that looks like tiny cream rocks and is great at retaining moisture; I like to sprinkle a fine layer onto the surface of my pots to improve germination
- **plant labels** – cut up a plastic milk bottle or yoghurt container, or use popsicle sticks (which can be composted); if you buy new plastic labels, then choose UV-resistant ones that won't become brittle and snap after one season
- **heat mat** – highly recommended if you live in a cooler climate like I do, as it makes germinating warm-season plants much quicker and easier
- **large plastic storage container with lid** – makes a great DIY greenhouse that can help maintain warm temperatures and reduce moisture evaporation
- **seaweed tonic** – stimulates better root growth, reduces transplant shock in young seedlings and helps plants through times of stress (and it can be used on all plants – young and old).

## HOW TO RAISE SEEDS

There are three stages when raising seeds: germinating, nurturing and hardening off. During seed germination, we wait for the seed shell to burst open. Here are the four things to consider when germinating seeds.

1. **Depth** – Sow small seeds on the surface of the seed-raising mix, then cover with a fine layer of vermiculite. For larger seeds (such as those of beans and pumpkin/squash), plant at a depth equal to twice their width. Don't sow seeds any deeper, or they might rot.
2. **Light conditions** – Most seeds prefer darkness to germinate, which is one of the reasons we bury them. However, a select few (such as shiso and celtuce seeds) prefer light. Sow these seeds on the surface of moist soil, and press down firmly so they make good contact with the soil.
3. **Temperature** – Different seeds require different soil temperatures to germinate, so check your seed packet. As a rule of thumb, cool-season seeds germinate in cooler temperatures, while warm-season seeds germinate in warmer temperatures.
4. **Water** – Always moisten the soil before sowing seeds, and gently water again after sowing. Keep seeds moist until they germinate by gently misting the soil with water on a regular basis. Some seeds (such as shiso seeds) need to be soaked in water prior to sowing.

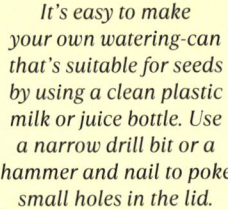

*It's easy to make your own watering-can that's suitable for seeds by using a clean plastic milk or juice bottle. Use a narrow drill bit or a hammer and nail to poke small holes in the lid.*

The moment your seedlings pop up from the soil, you need to start nurturing them. At this stage, your plant babies will require light, so move them to a bright spot by a window. The first two leaves you'll see are called cotyledons. These are the 'baby leaves' or 'nursery leaves' of the plant and are there to help grow the next set of leaves, known as the 'adult leaves' or 'true leaves'. Continue to keep the seedlings well-watered during this stage.

Your seedlings are ready to transplant into the garden once at least two sets of adult leaves have developed. If you raised your seedlings indoors, then you need to slowly acclimatise them to the great outdoors. This is called **hardening off** and involves gradually exposing them to outdoor conditions for a few hours a day, increasing exposure until they've completely adjusted so they don't suffer from transplant shock. An easy way to do this if you've used a plastic storage container as a greenhouse is to pop the container outside and keep the lid slightly ajar for the first few days, remove the lid for the next few days, then eventually take the seedlings out of the container.

### DIRECT SOWING

This follows the same process outlined for raising seeds in punnets, except that you're sowing seeds straight into the garden. Plants such as snow peas (mange touts) and long beans typically prefer direct sowing, as they are delicate and don't enjoy being transplanted. When direct sowing, it's vital to plant seeds at the right soil temperature. As a general rule of thumb, warm-season seeds require a soil temperature of 20–30 degrees Celsius (68–86 degrees Fahrenheit) for germination, while cool-season seeds require a soil temperature of 10–20 degrees Celsius (50–68 degrees Fahrenheit) for germination.

## PESTS AND DISEASES

Once your Asian veggie patch is underway, pests and diseases may find their way into your garden. Dealing with pests and diseases can be frustrating, but it's a shared challenge to which we can all relate.

Common pests you might come across include aphids, cabbage moth and cabbage butterfly caterpillars, slugs, snails, rodents, possums and birds. As for diseases, these are caused by fungi, bacteria and viruses and include powdery mildew, leaf spot, rust and blight.

My personal method for managing pests and diseases is to use a mix of gentle and preventive strategies, rather than hitting hard with harsh chemicals. This approach is referred to as integrated pest management, and it combines cultural, biological and mechanical solutions to control unwanted guests in the garden, so it's friendlier to the environment. Here are some strategies that you can try.

### GROW HEALTHY PLANTS

Like people, healthy plants are in a better position to withstand damage from pests and diseases. It's thought that some pests find it harder to digest healthy plants so are less likely to be attracted to them. Keep your plants healthy and free of stress by ensuring that they're adequately fed and watered, and protect them from extreme weather conditions such as drought or hot weather.

## LIVE IN HARMONY WITH NATURE

One of the best mindsets to have when gardening is to accept imperfection. Nothing is perfect, and it's difficult to control everything. We as humans are only one part of a wider ecosystem. Local wildlife, while frustrating to deal with, also needs food to eat. We did, after all, take over their natural habitats and food sources, building concrete suburbs on land that was once naturally wild. Accepting some pest damage is a gentle way of acknowledging the fact that we are part of something greater, and that we should live in harmony with nature and each other.

### ACT QUICKLY

Reduce the impact of many diseases by removing affected plant material as soon as you see it. Pests can be controlled easily in the same way. I've never seen anything reproduce as quickly as aphids — just knock them off with a spray of water. With slugs and snails, go outside at night with a head torch and pick them off with a pair of tweezers. It's quite satisfying to do!

### HAVE GOOD WATERING HABITS

Water the soil and not the plants, and do it in the morning instead of the evening. This will help to reduce the spread of fungal diseases, such as powdery mildew.

### IMPROVE VENTILATION

Fungal diseases are more likely to occur in damp, humid environments. Improve ventilation in your garden by leaving breathing space between plants. Prune leaves that touch the ground to stop soil-borne fungal diseases from creeping onto your plants. Keep fruit trees happy by removing crisscrossing branches and keeping the structure free of clutter — this ensures that air can circulate freely within the tree.

### ATTRACT BENEFICIAL PREDATORS

Plant flowers around your garden to create a wildlife-friendly environment that will attract beneficial predators (such as ladybirds, parasitic wasps, hoverflies and lacewings), which will keep pest populations under control. Avoid insecticides, which kill good bugs as well as bad.

### TRY OTHER VARIETIES

Look for disease-resistant forms, or choose coloured versions of plants (for example, purple and white varieties) that can be difficult for pests to spot.

### KEEP TOOLS CLEAN

Always sanitise your tools before pruning or cutting into plants to stop diseases from spreading between plants. A 70:30 ratio of methylated spirits to water in a small spray bottle works well.

*Dragonflies are welcome visitors to the garden, as they love to eat mosquitoes and cabbage butterflies.*

## DO NOTHING

Instead of applying sprays that might throw things out of whack, let natural balances come into play. Nature has a wonderful way of working itself out without human intervention. Leave the aphids in the garden, and you might just find that predatory insects (such as ladybirds) will arrive to sort them out.

## PHYSICALLY EXCLUDE PESTS

Cabbage butterflies, birds and possums can easily be kept at bay with plant netting. Build a netting frame for garden beds using poly pipe, or throw a net over fruit trees when they start to bear fruits. Rodents can be a bit trickier to manage, as they can burrow underground, squeeze into small spaces and bite through netting. Prevent rodents from accessing your veggies by covering your plants with a cage made from rodent-proof mesh. Be sure to bury the cage edges into the soil of your garden beds to stop the rodents from burrowing under them. Trapping is another option, and peanut butter on bread makes an attractive bait. Avoid using second-generation baits, which are highly toxic and dangerous for pets and native wildlife.

## USE COMPANION PLANTING

Growing a mix of different plants together can slow down pests and stop diseases from jumping easily across plants of the same variety. Cabbage moths and cabbage butterflies locate mustard family plants (such as bok choy and Chinese broccoli) by scent and leaf shape, so planting other aromatic plants (such as garlic or spring onion/scallion) or plants with interesting leaf shapes (such

as nasturtiums and calendulas) can confuse them and help to camouflage target plants.

## PLANT SACRIFICIAL CROPS

If you have larger pests (such as possums) visiting your yard via the local possum highway (your fence), consider planting a sacrificial crop or leaving food scraps along the route so they don't venture further into your garden.

## PLANT TRAP CROPS

Some plants are attractive but toxic to certain pests. Land cress is a trap crop for brown cabbage moths (not white cabbage butterflies), which seek out the plant and lay their eggs on it. The caterpillars that hatch then feed on the plant and die because of the saponin (a soap-like compound) in the leaves.

## KEEP THE GARDEN CLEAN

To stop diseases from spreading throughout your garden, remove infected plant material promptly. This material is best kept out of small-scale cold-composting systems, which don't produce enough heat to kill off diseases. Dispose of this material in your council green bin, so it can go through a hot-composting process.

## CHOOSE LOW-TOX ALTERNATIVES

If nothing else works against small sap-sucking insects (such as aphids, mealy bugs and thrips), then make a low-tox spray by diluting soap in water. Horticultural soap is best, although kitchen soap can also work in a pinch (just don't use it too often). Spray it directly onto the bugs every few days to keep things under control. For slugs and snails, try a yeast trap or elemental iron pellets that break down into iron for your soil (these are not to be confused with traditional metaldehyde slug pellets, which are highly toxic to native wildlife and pets). For diseases, organic copper fungicides can be used according to the product directions but only sparingly (to prevent copper build-up in the soil, which can be harmful to plants).

## CAN CROP ROTATION SOLVE YOUR PROBLEM?

Crop rotation involves moving veggies to different garden beds each year so that pests and soil-borne diseases found in one patch won't affect the same plants the following year. This practice also helps to maximise plant productivity, as the same soil is not constantly being depleted of the same nutrients.

Crop rotation comes from large-scale monoculture agriculture and has been practised by farmers for centuries, but it's unclear whether it helps (or is practical) in a smaller urban garden. This is because when you're rotating crops on a small scale, it's difficult to isolate pests and diseases simply by moving things a short distance away. Also, most small gardens contain a mix of different plants, rather than many of the same plants in the one spot, and this – combined with the addition of lots of organic matter – can help to build a resilient and biodiverse ecosystem. And it's difficult to rotate crops if you have only a few garden beds – you'll run out of space pretty much immediately.

I personally feel that, for urban gardens, it's more important to make sure that you're always replenishing the soil with organic matter, planting a diverse range of veggies and using other preventive approaches to pests and diseases. So, don't stress too much if you can't make crop rotation work in your space.

# Gardening in a small space

With so many plants to choose from, it's easy to get carried away and want to grow everything. Unfortunately, my garden size won't allow me to plant 150 different fruit trees (yes, my wish list is that long!). In the case of an urban garden, one of the most important things we need to do is to prioritise. Here are some ways to decide what to grow:

1. **Grow what you like to eat** – I love eating bok choy, snow peas and celtuce, so I grow lots of these. As fun as it is to grow some veggies (hello bitter melon and your beautiful fruits), if you don't like them all that much, or only plan to eat them once or twice a year, then it's probably not worth devoting garden space to them.
2. **Grow what's hard to buy** – Shops only have produce that is easy to transport, store and sell. Because of this, lots of Asian veggies aren't found in the shops. Where I live, it's hard to find celtuce, fresh jujubes and sweet potato leaves for sale, so I prioritise growing these in my garden.
3. **Grow quick-growing veggies** – Plants such as choy sum, bok choy and Asian mustards only 'rent' garden space for a short time. This means that you can plant multiple batches of these veggies in the same time it would take to cultivate a slow-growing plant such as garlic.
4. **Grow produce to share with your community** – If you have a gardening buddy who's keen to start their Asian veggie patch journey with you, then why not share the growing between the two gardens and meet up once a fortnight to swap harvests? Alternatively, look for local crop swaps, where gardeners get together once a month to exchange excess crops.

## MAKING THE MOST OF SMALL SPACES

Our world needs more people growing food plants in urban areas. This would transform the social, cultural and environmental landscape of our cities.

Growing in small urban spaces can help to greenify our concrete cities, bring back biodiversity and reconnect people with nature. Turning unused space into a place of food production helps to improve food security and build community resilience. Being close to nature can encourage people to live more sustainably and protect the future of our planet. All of these benefits have inspired me to make the most of my urban garden.

There's a lot of underused space around our homes and gardens that, if used efficiently, could provide more than enough food for a household's needs. Following are some tried and true methods of making the most of small spaces.

1. **Grow vertically** – If we visualise the 3D space in our garden, rather than just the surface area available, we will realise that we have a lot more space than we thought. Make the most of boundary fences, front fences, back fences, gates and sides of houses as vertical supports for climbing plants. Attach garden pots to a wall in order to grow dozens of plants in the space of one. Trail plants across arches, balcony or porch railings and even the top or sides of a carport.
2. **Grow on concrete** – Use pots or raised beds to turn empty concrete patios, driveways, balconies, stair edges or porches into spaces of abundance. To manage water drainage, use self-contained wicking beds and place saucers under pots. Concrete retains warmth, so it creates an ideal microclimate for heat-loving plants.
3. **Grow indoors** – Create an edible garden inside by swapping traditional house plants with veggies and herbs. Cultivate sprouts and microgreens on sunny windowsills, or use a shelf to grow them vertically in front of a large window. Grow mushrooms in your bathroom or on your kitchen bench. Raise an endless supply of mung bean sprouts in a jar – they don't even need light!

# Harvesting

Nothing beats the excitement of your first garden harvest, when the Asian veggies you've been carefully keeping tabs on are finally ready to be picked and cooked up into a garden-to-table homegrown meal – that is, if they make it to the kitchen (quite a few of my veggies are gobbled down raw while I'm still in the garden)! It's just like heading to an Asian grocer, only it's right at your doorstep - and I assure you that the veggies will be the finest ones you've ever tasted.

## ADVICE FOR HARVESTING

When it comes to harvesting your own produce, it can be difficult to know when things are ripe enough to pick, since farmers usually make this decision for us with store-bought veggies. Here are some general pointers on how to get the best and freshest harvests from your garden – you can find even more information in the harvesting section of each plant profile.

- **Early morning is the best time to harvest veggies** – This is when there is more moisture in them, so they'll taste better and last longer when stored.
- **Young produce tastes better** – As fun as it is to see how large melons, beans and mustard leaves can get, most of the time you'll find that the smaller they are, the better they taste. If you're not sure, then use the size of supermarket veggies as a guide to how big your harvested veggies should be. That said, experiment with harvesting at different stages, as the beauty of growing your own veggies is the freedom to harvest whenever you like, to be creative and to make the most of your produce. Extra young broad beans can be eaten as green beans; extra old loofahs can be transformed into shower sponges.
- **The more you harvest, the more you'll get** – Harvest plants as soon as they're ready. Doing so allows plants to direct their energy to growing more produce for you.
- **The more you harvest, the longer you can harvest** – If you don't harvest the produce that your plants are giving you (such as beans), then they'll get confused and focus on maturing the crop. This means that they will produce seeds earlier, resulting in a shorter harvest season. If you keep harvesting, then the plants will continue to focus on growing instead of maturing.
- **Keep the stems** – For veggies such as pumpkin (squash) and chilli, harvesting the produce with a bit of stem attached will help it last longer when stored.
- **Enjoy 'cut and come again' greens** – Many Asian greens do not need to be harvested whole, the way they're sold in supermarkets. You can harvest the outer leaves only, leaving the inner ones to keep growing. This gives you a longer and continual supply of greens.
- **Make sure that you wash your veggies** – While organically grown veggies won't be covered with pesticides, produce grown in urban areas can be contaminated with other things, such as dust or pollution. Washing with tap water is a simple and effective way to rinse your veggies. Some people add a dash of apple cider vinegar, white vinegar or salt to the water for peace of mind.

56

*Leafy greens with little holes caused by insects are usually safe to eat. Simply remove the damaged areas and consume the rest.*

*If the damage is caused by rodents or animals, then it's best to compost just in case. The plant material eventually returns to the ecosystem to help grow next season's veggies.*

*One of the greatest joys of gardening is being able to share your excess harvests with others. So, pop by your neighbour's or friend's home with a paper bag filled with freshly harvested veggies, or leave a basket in front of your home for passers-by.*

# Cooking

To make the most of your Asian veggie harvests, add the following Asian kitchen staples to your pantry – you'll find them featured in many of my Chinese recipes. I've also listed some helpful cooking tools, as well as storage ideas for keeping your harvests fresh.

## PANTRY INGREDIENTS

All of these ingredients are commonly used in Asian cooking. If you find that you have leftovers, then I'm confident you'll be able to use them in other Asian dishes. Most of these ingredients can be found in large supermarkets these days, but if you're having trouble finding something, then head to an Asian grocer – I've included the Mandarin name for the Asian ingredients, so you know what to look for.

### NEUTRAL OIL WITH A HIGH SMOKE POINT

This type of oil is perfect for a hot and sizzling stir-fry. I usually go for vegetable or grapeseed oil. You can also try another neutral oil, such as canola, sunflower or soybean. Peanut oil is a very popular option, but it can impart a (sometimes welcome!) nutty aroma.

### SESAME OIL (芝麻油)

The most divine ingredient, sesame oil is a seasoning oil rather than a cooking oil, and it adds a nutty, sesame touch to dishes. You either use it as a garnish or add it at the end of the cooking process. If it's cooked for too long, it loses its aroma. I should say that the oil to which I am referring is toasted sesame oil, which is dark brown in colour and often labelled as 'sesame oil'. You can sometimes find untoasted sesame oil, which is golden in colour and not the same thing.

### CHILLI OIL (辣椒油)

If you love chilli, then you've absolutely got to add chilli oil to your pantry. It's oil infused with chilli, but you can also find

it with additional spices, aromatics or even peanuts. While you can buy it, you can also make it easily at home (see page 135).

### LIGHT SOY SAUCE (生抽)

Probably the most widely known Asian cooking ingredient, it is used in many of my recipes. Light soy sauce adds a savoury umami flavour to dishes.

### DARK SOY SAUCE (老抽)

This sauce is thicker than light soy sauce. It adds a rich caramel colour and a slightly sweet taste to dishes.

### OYSTER SAUCE (蚝油)

It's made from 'oyster essence' and has a thick, sauce-like consistency. The sweet, salty, umami flavour pairs deliciously with Asian leafy greens.

### HOISIN SAUCE (海鲜酱)

Hoisin sauce is made from fermented soybean paste, and it has a similar sweet, salty, umami flavour. You'll find that it's sometimes used as a vegan substitute for oyster sauce, although it doesn't taste completely the same, as it's much sweeter. This is a key ingredient in my much-loved hotpot dipping sauce (see page 98).

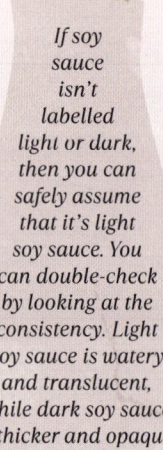

*If soy sauce isn't labelled light or dark, then you can safely assume that it's light soy sauce. You can double-check by looking at the consistency. Light soy sauce is watery and translucent, while dark soy sauce is thicker and opaque.*

### CHINESE BLACK VINEGAR (CHINKIANG VINEGAR) (镇江香醋)

Made from fermented glutinous rice, black vinegar is a combination of sweet, salty, fruity and umami flavours. It's somewhat similar to balsamic vinegar but less sweet. It's a must-have ingredient for making my favourite dumpling dipping sauce (see page 111). There are four main types of Chinese black vinegar, from different regions in China, but the one I use (and probably the best-known type) is called Chinkiang vinegar.

### RICE VINEGAR (RICE WINE VINEGAR) (米酒醋)

Used in Japanese cooking, rice vinegar has a light vinegar taste. I love using it as a base for Asian-style pickles, and it also gives sushi rice its delicious flavour.

### CHINESE COOKING WINE (SHAOXING WINE) (中国料酒)

This wine is made from glutinous rice and has a light, sweet, vinegar flavour. While the flavour is hard to distinguish, it can transform a dish from pretty good to incredible. Kind of like how white wine is used in cooking – it just gives dishes that extra oomph.

### CHICKEN BOUILLON POWDER (鸡粉)

Basically dehydrated chicken broth, this powder is used to add a umami flavour to dishes. It's a cooking staple and can be added to pretty much anything. There are vegetarian versions, too.

### SICHUAN PEPPERCORNS (四川花椒)

These are the berries of the prickly ash tree, and they give dishes a spicy, citrusy taste. Sichuan peppercorns can make your mouth feel numb, so they're usually added at the start of the cooking process to flavour the oil, then removed.

### GROUND WHITE PEPPER

This is the go-to ingredient for adding peppery heat to Asian dishes. To me, it tastes stronger and spicier than black pepper, but blends in well since the particles are so fine.

### COOKING SALT

All recipes in this book use cooking salt unless another type is specified. You can replace cooking salt with table salt, but make sure that you use less – table-salt granules are smaller, so there are more of them in each teaspoon.

## SHIITAKE MUSHROOMS
(香菇)

My go-to mushrooms for Asian cooking, shiitake mushrooms add an aromatic, earthy, umami flavour to dishes. I also love using them as a plant-based substitute for meat. (If you don't have shiitake mushrooms, then you can replace them with other Asian mushrooms, such as enoki, oyster or shimeji mushrooms.)

For most of my cooking, I use dried shiitake mushrooms after rehydrating them. Dried shiitake mushrooms have a much stronger umami flavour than fresh shiitake mushrooms, and they're cheaper. Buy a bulk bag of dried shiitake mushrooms, and split them between friends!

Be sure to rehydrate them for six to eight hours – I usually soak them the night before or on the morning I plan to use them.

Rehydrate them in a bowl of room-temperature water, and pop a small plate on top to keep them submerged. Once rehydrated, rinse the mushrooms to remove any dirt.

## BLACK WOOD EARS (黑木耳)

Delicious wild fungi that grow on trees, black wood ears have a light, woody taste and are prized for their incredible texture. They give a crunchy, chewy texture to stir-fries, but when cooked for a long time in soups, they transform to have a soft, jelly-like texture. Black wood ears can sometimes be found fresh, but they're more commonly available dried. Like shiitakes, they need to be rehydrated before cooking. They are much thinner than shiitake mushrooms, though, so only need one to two hours of soaking. A small handful goes a long way, as they expand to five times their dried size.

*Rehydrating shiitake mushrooms in room-temperature water makes them taste better, but it takes time. If you forget to rehydrate your mushrooms before you're ready to cook with them (as I often do!), then you can quickly boil them in hot water for 30 minutes.*

## STORAGE ESSENTIALS

Growing your own Asian veggies means that you're not buying veggies with single-use packaging. But without the packaging, it can be a challenge to figure out how to store your homegrown harvests. Over the years, I've come up with a list of essential re-usable items that will allow you to store your veggies in an eco-friendly way that keeps them fresh.

- **Large, flat, rectangular cupcake boxes** are amazing for storing Asian greens and anything with long, flat leaves.
- **Cake boxes** are great for mini womboks as well as pumpkins (squashes) cut in half.
- **Plates** can act as plastic wrap for cut pumpkins (squashes) and melons – just pop them on with the cut side downward.
- **Fabric napkins and linen towels** are handy for wrapping bunches of veggies to keep them fresh in containers.

- **Re-usable lunchboxes** are useful for storing little things such as ginger rhizomes, as well as for freezing portions.
- **Glass jars** can be filled with water to keep herbs fresh, or used for making fridge pickles. Re-use supermarket glass jars by removing the labels in hot water – you can also buy new lids for them if the old ones are smelly.
- **Re-usable silicon bags** are good for portioning and freezing beans and chopped veggies.

# PART TWO

## Plant

# PROFILES

In this section, you'll find a carefully curated list of 40+ of my favourite Asian veggies, herbs and fruits to grow at home. I've chosen each of these based on their suitability for backyard growing, and focused on plants that are easily cultivated, grow quickly or deserve a place in smaller or urban gardens. For those who love to experiment (like me!), there are also some fun and challenging plants included that are so worth giving a go.

There is a good balance of annuals and perennials, and many plants suit a broad range of climates. The veggie section has been divided into cool-season and warm-season plants, to make it easier to plan out your gardening schedule.

Much of what is within these pages is inspired by my own lived experience and culture. It is by no means a representation of every Asian culture – Asia is a continent of countries that are so diverse, this richness and complexity could not fit into one book. As a nod to my Chinese heritage, I've included the Mandarin names for plants, along with the pinyin (Romanised version), should you wish to immerse yourself in the experience and learn a bit of my family's native language as you read along.

# COOL-SEASON

# VEGGIES

# ASIAN MUSTARDS

*Brassica juncea*
Mandarin: 芥菜 – jiè cài
Chinese mustard • gai choi • gai choy • leaf mustard • oriental mustard

| | | |
|---|---|---|
| 🏷 | **PLANT TYPE** | Cool-season annual |
| 🌰 | **PLANT FAMILY** | Mustard family (Brassicaceae) |
| ⤢ | **PLANT SIZE** | 30–40 cm (12–16 in) tall, depending on variety |
| 🪴 | **POT FRIENDLY?** | Yes, plant one in a 30 cm (12 in) pot or a few in a larger pot |
| ✺ | **SUN** | Full sun or part-sun |
| ❄ | **FROST** | Frost tolerant |
| 💧 | **WATER** | Regular watering |
| ⬡ | **FOOD** | Medium feeder |

Asian mustards are a diverse and interesting group of veggies commonly used in stir-fries and pickle-making; tender and leafy varieties are used in stir-fries and as baby salad greens, while strong, spicy and pungent varieties are typically reserved for pickles. They're easy-to-grow, compact plants, just like any other leafy greens, but their peppery punch provides a welcome contrast in flavour.

I love growing Asian mustards for homemade Chinese pickles, which can be eaten as a side or appetiser, or used as a cooking ingredient to add a salty, umami taste to stir-fries and other dishes. Chinese pickles have a special place in my heart, as they are a part of my family's cultural heritage. When I was a kid, my mum always had little packets of Chinese pickles sitting in the fridge. She would snack on them or use them in her cooking, as they reminded her of her homeland.

# VARIETIES

I have never seen a family of Asian greens as diverse as the mustard family. I'd like to introduce you to three types: thick-stalked mustards, swollen-stem mustard and leafy green mustards. Each is used to make a different kind of traditional Chinese pickle.

**Thick-stalked mustards –** These are used to make one of my favourite Chinese pickles, suan cai (see page 74). It's a sweet-and-sour pickle that is often known as Chinese sauerkraut. The most traditionally used mustard is **gai choi**, a popular heirloom form that looks like a loosely wrapped cabbage when mature. It has a moderate peppery flavour.

Any other similar-looking mustards also work, as long as they have a thick, meaty stalk because that's the best part to pickle. I love trying different varieties and personally like to grow a Japanese one called **Red Giant**, as I'm a big fan of its reddish purple leaves.

**Swollen-stem mustard –** Part of the *Brassica juncea* Tsatsai group and originating from the Sichuan province in China, swollen-stem mustard is a distinctive plant that grows a knuckle-like stalk. Think of it as an Asian version of kohlrabi.

Swollen-stem mustard is used to make zha cai (see page 73), China's most popular pickle. The Chinese government once used zha cai sales around the country to measure inter-province population migration. If sales went down in one city, then it suggested that fewer people were living there; if sales went up, then it suggested that many migrants were moving in.

**Leafy green mustards –** Tender and leafy varieties are used to make a Chinese pickle called xian cai (see page 73). This pickle is extremely popular in Shanghai, where my parents are from. It is traditionally made with a specific mustard called **xuelihong**, which is difficult to locate outside China.

However, I find that any leafy varieties of mustard can be used to make xian cai. I've had great success with **Chinese mustard greens**. These foliage-rich mustards can also be stir-fried as leafy greens.

## TYPES OF ASIAN MUSTARDS

*Thick-stalked*    *Swollen-stem*    *Leafy green*

# GROWING

Asian mustard seeds are small, so sow them on the surface and cover them with a layer of vermiculite. Water in gently. Either raise seeds in punnets or direct sow them in the ground, then thin them out. Seeds germinate at around the one-week mark. Transplant seedlings into the garden once three or four adult leaves have appeared.

Asian mustard plants prefer a full-sun position in cooler climates and a part-sun position in warmer climates. They're fast growers when temperatures are cool. Asian mustard plants prefer well-drained soil but are tolerant of poorer soils.

Larger varieties (such as Red Giant) require more room, so allow 40 centimetres (16 inches) of space around each plant. More compact, leafy mustards require less room – 30 centimetres (12 inches) of space around the plant is sufficient.

Asian mustard plants taste best when they grow fast, so give them plenty of water throughout the growing season. Under-watering can cause leaves to turn bitter, while watering them inconsistently can lead to early bolting. Interestingly, the leaves become sweeter when the plant experiences a period of light frost, and they are spicier when the plant is grown during warm weather.

Keep an eye out for white leaf spot, as plants are susceptible to this fungal disease. It shows up as little white spots on older leaves. Remove affected leaves as soon as you see them to prevent the disease from spreading.

# HARVESTING

It usually takes about two months for **thick-stalked mustards** to be ready to harvest whole. But you can also treat them as 'cut and come again' veggies and harvest only the outermost leaves while they're growing.

**Swollen-stem mustard** is slow growing and takes three to four months to be ready for harvesting. This one is harvested whole, as you grow it for its stem.

**Leafy green mustards** can be harvested continually over the season, using the 'cut and come again' method, until the plant starts to bolt. The larger and more mature the mustard leaves, the more pungent they taste. If you're planning to make pickles with them, then it doesn't matter as much. But if you want to cook with them, harvest them young.

## NATURE'S CLEANSING GREENS

*Brassica juncea* mustards are often included in biofumigant green-manure mixes, as they help to reduce soil-borne fungal diseases and harmful nematodes (such as root-knot nematode). When growing mustards as a green manure, let them mature until just before they flower, then cut back the plants at the base and chop them into smaller pieces. Leave the pieces on the surface of the soil or gently dig them in. As the plants break down, they release their fumigation magic into the soil. The decomposing plants also add bulk organic matter to the soil.

*If you direct sow your seeds, then thin out plants and use the unwanted seedlings as tasty microgreens in sandwiches and salads. Young mustard leaves are especially tender and mild in flavour.*

## STORING

Store thick-stalked mustards and leafy green mustards unwashed in a re-usable container in the fridge, with a cloth napkin to absorb moisture. For swollen-stem mustard, remove the leaves and store them in the same way in the fridge. They should all keep fresh for up to a week.

## COOKING

Turn to pages 73–4 to discover four fabulous recipes for Chinese pickles made from Asian mustards. Once you've made a batch, try the following two recipes to see how easily the pickles can be used in cooking or as a porridge topping. Feel free to experiment as well. Although these two recipes call for suan cai, any Chinese pickles can be substituted to add their own unique flavour.

### Edamame and pickled mustard greens

Who knew that such a simple pairing would be packed with so much flavour? This is a family favourite of ours. Enjoy it as a veggie side when freshly cooked, and combine any leftovers with rice to make quick and easy fried rice the next day. It can be eaten hot or cold, and it's especially refreshing as a cool dish in summer.

Prepare the suan cai by first gently squeezing out any excess liquid. Then finely chop and set aside. Finely chop the garlic cloves. Chop the chilli (if using).

Heat the vegetable oil in a wok over medium heat. Add the garlic, and stir-fry until fragrant. Add the edamame beans, and stir-fry for 2–3 minutes.

Stir in the cooking salt and water. Cover the wok with a lid, and allow it to simmer for 5–6 minutes or until most of the water has evaporated.

Add the suan cai, light soy sauce and chilli (if using). Stir to combine, then serve.

**SERVES 4**

150 g (5¼ oz) suan cai (or other Chinese pickles; see page 72)

2 garlic cloves

1 fresh chilli (optional)

1 tablespoon vegetable oil

350 g (12¼ oz) edamame beans (if using frozen, thaw and rinse before using)

1 teaspoon cooking salt

½ cup (125 ml) water

1 teaspoon light soy sauce

## Pork floss and egg congee with Chinese pickles

Chinese rice porridge (congee) is a childhood favourite of mine. It's such a comfort food and makes for a fabulous Asian-style breakfast or brunch at home. A spoonful of Chinese pickles is the perfect congee topping and adds a big burst of flavour.

This is a quick go-to recipe of mine using pork floss and egg – two things we always have at home. You can experiment by swapping pork floss for other savoury ingredients (such as dried scallops or shrimps), and each soft-boiled egg for a century egg. Vegetarians and vegans can replace the pork floss with mushroom/vegetarian pork floss, Chinese doughnut stick or tofu.

Rinse the rice and drain. In a bowl, prepare a slurry by mixing the cornflour and cold water.

For the topping, soft-boil the eggs, and cut them in half. Finely chop the Chinese pickles and the spring onions.

Place the rice and broth into a saucepan, and bring to a boil. Reduce the heat to low, and simmer for 30 minutes.

Once the rice has finished simmering, stir in the cornflour slurry, light soy sauce, cooking salt and ground white pepper. Cook for a further 1 minute.

Divide the congee into 2 large bowls, then top each with 2 egg halves, 1 heaped tablespoon Chinese pickles, spring onion and 2 tablespoons pork floss. Serve hot.

**SERVES 2**

**For the congee**

½ cup (110 g) jasmine rice

1 teaspoon cornflour (cornstarch)

2 teaspoons cold water

4 cups (1 litre) chicken, pork or veggie broth

½ teaspoon light soy sauce

½ teaspoon cooking salt

¼ teaspoon ground white pepper

**For the topping**

2 eggs

2 heaped tablespoons Chinese pickles (see page 72)

2 spring onions (scallions)

4 tablespoons pork floss

# Chinese pickles

Making pickles is the perfect activity for peak harvest time in the garden, when you have an overabundance of veggies. It helps to preserve excess crops and stop them from going to waste, and it spreads out the gluts so you can enjoy them during the quieter times in the garden.

There are so many ways you can use pickles: as an appetiser or side, on porridge, in stir-fries or fried rice ... just to name a few. Here you'll find four quick and easy recipes to get you started on your pickling journey. Feel free to scale the recipes up or down, but ensure that you stick to the same ratios. These pickles should last in the fridge for up to three weeks but will probably be eaten before then. Be sure to always use a clean spoon when scooping them out. Discard the pickles if you see any signs of mould.

## Xian cai (咸菜)

Xian cai is salted mustard greens (*xian* means salty, while *cai* means veggies), and it is best eaten with other dishes. We always have some in our fridge, ready to go. It's great with rice or porridge, and it gives stir-fries a salty kick. To store the pickle, you'll need a clean and sterilised glass jar with a lid (see opposite).

Wash the Asian mustard leaves, and shake them to get rid of excess water (they don't need to be completely dry). Cut into 1 centimetre (½ inch) wide pieces, and place into a large bowl. Massage the sea salt into the leaves until the leaves start to bruise and liquid comes out.

Chop the chillies (if using). Add the chillies, sugar and vegetable oil to the bowl with the Asian mustard leaves and liquid, and mix to combine.

Transfer the mixture to the jar, and use a spoon to push down the Asian mustard leaves so they're completely submerged. Place the lid on the jar.

Set the pickle aside in the fridge for 1–2 days before eating.

**MAKES 2 CUPS (500 G)**

350 g (12¼ oz) Asian mustard leaves (leafy green variety)

2 teaspoons sea salt

2 fresh chillies (optional)

1½ teaspoons sugar

2 teaspoons vegetable oil

## Zha cai (榨菜)

Zha cai is crispy and salty with a hint of sugar and spice. It can be used in numerous ways, but my favourite is to add it to my two-minute silken tofu (see page 252). Traditional zha cai takes months to prepare, so this recipe is a modern version that you can make quickly at home. You'll need to salt the mustard stem for 2 hours, so allow time for that. To store the pickle, you'll need a clean and sterilised glass jar with a lid (see opposite). Zha cai can also be made using kohlrabi.

Wash the mustard stem, and peel the skin. Slice into matchsticks, and place into a bowl. Massage the sea salt into the stem pieces, and set aside for 2 hours. The stem pieces will soften, and water will be drawn out.

Finely chop the garlic clove and ginger, and chop the chilli (if using).

Discard the water, rinse the stem pieces and place them into another bowl. Add the rest of the ingredients, and mix together.

Transfer the mixture to the jar, and place the lid on. Set the pickle aside in the fridge for 1–2 days before eating.

**MAKES 1 CUP (250 G)**

250 g (9 oz) swollen-stem mustard (stem)

2 teaspoons sea salt

1 garlic clove

1 cm (½ in) piece of ginger

1 fresh chilli (optional)

1 tablespoon chilli oil (see page 135)

1 tablespoon sugar

2 teaspoons white wine vinegar

1 teaspoon light soy sauce

½ teaspoon fennel seeds

1 star anise

# Pickled celtuce stem (醬菜心)

You might have seen this in Asian grocers labelled as 'pickled lettuce' but noticed that the jar is filled with round cucumber-like discs that look nothing like lettuce. I, too, wondered how it got lost in translation. Then one day I came across a brand with an illustration of celtuce on the label – and the mystery was solved. I guess they weren't wrong – celtuce is, after all, a type of lettuce! After a bit of reverse engineering, I managed to figure out how to make it myself at home. Here, celtuce stems are soaked in a soy brine, which is a popular way of making pickles in Asia. To store the pickle, you'll need a clean and sterilised glass jar with a lid (see page 72).

Peel the celtuce stem (see page 88), and slice it into thin pieces. Place the pieces into a bowl. Add the sea salt and ½ teaspoon sugar, and mix well. Set aside for 1 hour to draw out the water. Discard the water.

Add the light soy sauce, rice vinegar and 3 teaspoons sugar to the jar. Mix well until the sugar has dissolved (heat it a little in the microwave if necessary).

Transfer the celtuce pieces to the jar, and use a spoon to push them down so they're completely submerged. Place the lid on the jar.

Set the pickle aside in the fridge for 1–2 days before eating.

**MAKES 1 CUP (250 G)**

350 g (12¼ oz) celtuce (approx. 1 stem with attached leaves), to yield 150 g (5¼ oz) peeled celtuce stem

½ teaspoon sea salt

3½ teaspoons sugar

1½ tablespoons light soy sauce

2 teaspoons rice vinegar

# Suan cai (酸菜)

Soft, sweet and sour, suan cai is traditionally preserved by lactic acid fermentation, but here we're doing a quick refrigerated-pickles version. You'll need to salt the wombok for 1 hour, so allow time for that. To store the pickle, you'll need a clean and sterilised glass jar with a lid (see page 72). Suan cai can also be made using thick-stalked Asian mustard varieties (see page 67).

Cut the wombok into 2 centimetre (¾ inch) wide pieces, and place into a large bowl. Massage the sea salt into the wombok pieces, and set aside for 1 hour to draw out the water.

Chop the chilli (if using). Add the chilli and Sichuan peppercorns to the glass jar.

Squeeze out as much water as possible from the wombok pieces, and tightly pack them into the jar.

Stir the vinegar and sugar together in a cup until the sugar has dissolved (heat it a little in the microwave if necessary). Pour this mixture into the jar, and use a spoon to push down the wombok pieces so they're completely submerged. Place the lid on the jar.

Set the pickle aside in the fridge for 1–2 days before eating.

**MAKES 2 CUPS (500 G)**

300 g (10½ oz) wombok

1 teaspoon sea salt

1 fresh chilli (optional)

¼ teaspoon Sichuan peppercorns

3 tablespoons rice vinegar (or white vinegar)

1½ tablespoons sugar

*Clockwise from top left:*
*suan cai; zha cai; xian cai;*
*pickled celtuce stem*

# BOK CHOY

*Brassica rapa* var. *chinensis*
Mandarin: 青菜 – qīng cài
bok choi • pak choi • pak choy • xiao bai cai

| | | |
|---|---|---|
| 🌱 | **PLANT TYPE** | Cool-season biennial, usually grown as an annual |
| 🌰 | **PLANT FAMILY** | Mustard family (Brassicaceae) |
| ⤡ | **PLANT SIZE** | Up to 30 cm (12 in) tall, up to 30 cm (12 in) wide |
| 🪴 | **POT FRIENDLY?** | Yes, plant one in a 25 cm (10 in) pot or a few in a larger pot or rectangle planter |
| ☀ | **SUN** | Full sun or part-sun |
| ❄ | **FROST** | Frost tolerant |
| 💧 | **WATER** | Regular watering |
| ⁙ | **FOOD** | Medium feeder |

Bok choy is a popular Asian veggie enjoyed outside of Asia, and it's often someone's first introduction to Asian greens because it's easy to find in supermarkets. I love growing it, though, because homegrown bok choy is so much fresher, and you can cultivate all sorts of varieties - with subtle differences in flavour - that you won't see in the shops.

Bok choy plants are compact and upright, and they take up barely any space in the garden. They also grow quickly and can be ready to pick within two months; baby greens can be harvested earlier. They're easy to look after and will happily grow in pots, making them great for balconies and small gardens.

Bok choy has a light and mild flavour, with tender leaves and crisp stems. It's a veggie staple in soups, stir-fries, noodle dishes, fried rice and curries. I can't think of a more versatile veggie, and I always make sure that I allocate a spot for it in my patch.

# VARIETIES

There are many interesting varieties of bok choy, but these are a few of my personal favourites:

- **Shanghai bok choy** – This is the most common variety in Asia, and the one sold in bunches at Asian grocers. It's sometimes called baby bok choy, green bok choy or pak choy, depending on the seed supplier. Shanghai bok choy is a small variety with light green stems that grows around 20 centimetres (8 inches) tall. Sweet and tender, it's less crispy than other varieties. It's the bok choy I grew up eating and the one I usually see in restaurants.
- **Purple** – An absolutely stunning ornamental-looking bok choy, it has vibrant purple leaves that turn green once cooked. I still love to grow it, nonetheless!
- **White-stemmed types** – These are larger than (and are not as popular in Asia as) green-stemmed Shanghai bok choy. However, they're sweeter and crisper than Shanghai bok choy, and taste a bit more watery, too.
- **Nai bai** – Also known as milk cabbage, nai you bai and nai you bai cai, this sweet and crunchy mini variety of bok choy grows up to 15 centimetres (6 inches) in height and has milky white stems and dark, crinkled leaves. Because it's small, all you need to do is chop it in half before cooking.

## BOK CHOY OR PAK CHOY?

These two names are often used interchangeably and refer to the same plant botanically: *Brassica rapa* var. *chinensis*. Pak choy is preferred in the United Kingdom and South Africa, while bok choy is used in the United States and Canada.

Where I live in Australia, the naming is not so clear. Asian grocers, my friends with Asian backgrounds and I know the green-stemmed variety as bok choy and the white-stemmed variety as pak choy. However, the NSW Department of Primary Industries defines them the other way around. When shopping for seeds, use the image on the packet to guide you, rather than choosing a variety because of its name.

# GROWING

Because bok choy is such a popular Asian green, you'll likely find seedlings available at the nursery. A punnet usually contains six to eight plants, which is an ideal number to start off with.

However, it's just as easy to grow them from seed. Sow seeds on the surface of the growing medium, and cover with vermiculite. Water in gently. The seeds will germinate in six to ten days.

Transplant the seedlings after three or four adult leaves have formed. Space the plants 20–30 centimetres (8–12 inches) apart. You can also direct sow the seeds densely, then thin the seedlings to 20–30 centimetres (8–12 inches) apart. Eat the unwanted seedlings raw as baby salad greens.

Bok choy plants prefer cool temperatures to grow, and taste especially sweet if allowed to grow through a period of cold weather. If you live in a warm climate, plant bok choy in part-sun and keep it well-watered. Heat and dehydration can increase the risk of bolting.

Ensure your soil is rich in organic matter. The faster the plant grows, the better it tastes.

Keep an eye out for white leaf spot, as bok choy is susceptible to this fungal disease. Remove affected leaves immediately.

## HARVESTING

The beauty of bok choy is that it can be harvested at any stage of growth. Baby bok choy plants around 10 centimetres (4 inches) in height are especially tender and delicious, so they are perfect for salads. I once saw a bowl of baby bok choy sitting next to baby spinach and rocket (arugula), served as a salad green at a hotel buffet in Singapore. It was delicious!

For larger leaves, you can harvest using the 'cut and come again' method, which allows you to enjoy a longer, continual harvest. Or wait until the plant has grown to the size you see in supermarkets, then harvest the entire plant by cutting it at soil level.

## STORING

Store your harvest unwashed, in a re-usable container with a cloth napkin to absorb moisture. Homegrown bok choy lasts much longer in the fridge than store-bought bok choy. I can store freshly picked bok choy in the fridge for more than a week, whereas supermarket bunches go limp within a few days.

## COOKING

Bok choy is such a versatile Asian veggie in the kitchen. It can be stir-fried with a variety of other ingredients, or on its own. My favourite combination is bok choy with chopped garlic and a pinch of salt, which is quite a simple dish. The recipe on the opposite page is a little more fancy!

78

*As they're compact, upright and quick-growing, bok choy plants are perfect for the gaps between larger, slower-growing cauliflowers and cabbages.*

*The bok choy plants will be harvested before your cauliflowers and cabbages start to take over the space. During summer, I like to pop bok choy seedlings under the shade of my tomatoes.*

*In addition, bok choy plants can be grown indoors as microgreens. They'll be ready to harvest in 10–14 days.*

### HOW TO PREPARE YOUR BOK CHOY

If you harvest your bok choy whole, then separate the leaves and wash each one in water to remove any soil at the base of the stalk. The entire plant can be eaten, including the middle 'heart', which is tender and sweet, and my favourite part of the bok choy. My sister discards this when cooking – she doesn't know what she's missing out on!

# Shanghai cai fan

*Cai fan* translates to 'vegetable rice' and is a fried-rice dish where the rice is infused with the flavours of the other ingredients as it cooks. It can be made with bok choy or any other Asian greens. I add Chinese sausage to my cai fan, but you can also choose bacon, kransky, five spice tofu or any flavoured firm tofu. This recipe uses a rice cooker, but you can also make it on the stove - just add more water.

Chop the Chinese sausages into small pieces. Dice the shiitake mushrooms.

Rinse the jasmine rice, and transfer it into a rice cooker. Add water as per your rice cooker's instructions, but keep it slightly on the dry side. I use 1½ cups (375 millilitres) water for 1½ cups (330 grams) rinsed jasmine rice. Set the rice aside – don't cook it yet.

Heat 1 teaspoon vegetable oil in a wok over medium heat. Add the shiitake mushrooms, and stir-fry for 1 minute or until fragrant. Add the Chinese sausage pieces and light soy sauce. Stir-fry for 1 minute.

Pour the wok mixture into the rice cooker. Add ¼ teaspoon cooking salt and the chicken bouillon powder (if using). Mix together.

Turn on your rice cooker, and let it cook as usual. While the rice is cooking, dice your bok choy and set it aside.

Once the rice is cooked, heat 2 teaspoons vegetable oil in a wok over medium heat. Add the diced bok choy and ¼ teaspoon cooking salt. Stir-fry for 2–3 minutes or until lightly cooked. Stir in the ground white pepper, then remove from the heat.

Transfer the contents of the rice cooker to the wok. Mix together and serve.

**SERVES 3–4**

120 g (4¼ oz) Chinese-style dried pork sausage (approx. 4 small ones)

4 large (or 5 small) shiitake mushrooms

1½ cups (330 g) jasmine rice

3 teaspoons vegetable oil

2 teaspoons light soy sauce

½ teaspoon cooking salt

¼ teaspoon chicken bouillon powder (optional)

350 g (12¼ oz) bok choy (approx. 2 big heads)

1 pinch of ground white pepper

## MORE WAYS TO USE BOK CHOY

Bok choy is a highly versatile veggie and can replace many other greens in curries, stir-fries, soups or stews. But if you ever have just a small harvest – not enough for an entire meal – then why not consider the following ideas?

- Chop and add it as a bonus veggie to instant noodles.
- Dice and add it as a bonus veggie to fried rice.
- Slice it into any fried noodle dish.
- Cut it up and add it to ramen, bone broth or any other soup-based dish.
- Combine it with other leftover veggies to make a soup stock.

# BROAD BEAN

*Vicia faba*
Mandarin: 蚕豆 – cán dòu
faba bean • fava bean

| | | |
|---|---|---|
| | PLANT TYPE | Cool-season annual |
| | PLANT FAMILY | Legume family (Fabaceae) |
| | PLANT SIZE | 60–180 cm (24–71 in) tall, 30–40 cm (12–16 in) wide; may need staking |
| | POT FRIENDLY? | Yes, grow multiple plants in a large pot |
| | SUN | Full sun or part-sun |
| | FROST | Frost hardy |
| | WATER | Regular watering |
| | FOOD | Light feeder |

Whenever broad bean season comes around in late spring, I eat these veggies by the bucketful. It's the only time I get to consume them fresh, as they're rarely available in the supermarket. They're deliciously tender, with a mellow, buttery, nutty taste.

Like all legume crops, broad beans are both multipurpose plants and easy to grow. In early spring, when pollen and nectar are scarce, their flowers are a valuable food source for bees. From late spring, they reward you with basketfuls of beans at a time when there's not much else to harvest. Then at the end of the season, they can be chopped off at the base to return nitrogen and bulk organic matter back to your soil. Broad beans are a must-have in any permaculture garden.

## VARIETIES

**Aquadulce** is a popular and prolific heirloom variety with extra-large pods up to 25 centimetres (10 inches) long. The plant grows to 1.5 metres (5 feet) tall. **Coles Dwarf** is an ancient, early-season variety growing only 1 metre (3 feet) in height. A prolific producer, its shorter height makes it easier to stake. **Crimson Flowered** is an heirloom variety with striking crimson blooms (most varieties have black-and-white flowers). However, in return for the unique flowers, you'll have to accept smaller (although still delicious!) beans. It's another short variety, growing to 1 metre (3 feet).

## GROWING

Broad beans prefer a long, cool growing season and won't flower properly in warm weather. If you live in a warm climate, double-check with local gardeners to see if you can grow them in your area, and plant accordingly.

Soak your broad bean seeds overnight before planting, and sow them 3–4 centimetres (1–1½ inches) deep. While they can be started in seedling trays, they're best direct sown. When direct sowing, space plants a minimum of 30 centimetres (12 inches) apart. If starting broad beans in seedling trays, then transplant them when two sets of adult leaves have appeared.

Broad beans prefer well-drained soil but tolerate heavy soil better than other plants. It's best to provide some support, as they grow quite tall and can bend in the wind. A great idea is to grow all of your plants close together, put some stakes around the perimeter, and secure the outer plants to the stakes with twine. The plants in the middle can lean on each other for support.

Keep an eye out for aphids (which love the tender growing tips) and slugs (which eat the broad bean seedlings).

### BROAD BEANS AS A GREEN MANURE

Grow broad beans in a dense group to use them as a green manure. They work especially well for clay soil, as broad beans are more tolerant of heavy soil than other plants, and the resulting organic matter helps to break up the clay clumps. If you want to plant broad beans en masse, then it's cost-effective to buy organic dried broad beans in bulk from an Asian grocer.

**Pruning** Broad beans grow more slowly than other beans, and it can be four or five months before they start to flower. Once the first flowers bloom (starting from the bottom of the plant), pinch off the growing tips. This tells the plant to stop putting on more leafy growth, and to start focusing its energy on producing flowers, leading to more broad beans to harvest.

The young growing tips can be eaten and have a mild flavour that resembles the beans themselves. Munch on them as a garden snack, or add them to whatever you're making for dinner that night, whether that's a stir-fry, soup or salad. In fact, you can grow broad bean shoots indoors as well (see page 208).

## HARVESTING

Broad beans can be harvested and eaten at various stages of growth. My personal preference is to harvest them when the beans have swelled large in the pods. Broad beans picked at this stage need to be double-shelled. However, broad beans can also be picked earlier and, if so, only need the outer pods removed.

Very young broad beans, around 5 centimetres (2 inches) long, can be picked and eaten as green beans. Unless I'm craving fresh green beans at that time of the year, I personally like to wait for the pods to grow bigger.

You can also leave broad beans to dry out on the plant and then use the dried beans in various dishes or save them as seeds for future sowings.

## STORING

If you want to save some of your fresh broad bean harvest for later, then freeze the blanched and double-shelled beans. I recommend freezing them in 400–500 gram (14–18 ounce) portions, so you can easily defrost a batch whenever it's needed.

## COOKING

How you cook broad beans depends on the stage at which you harvest them. With very young broad beans, I love to make a garlicky stir-fry (see page 161; replace the long beans or flat beans with young broad beans). For larger broad beans, I stick to simpler stir-fries. The buttery taste of broad beans goes well with salty flavours, so the beans pair beautifully with seafood. Boiled broad beans are also delicious and can balance strong flavours such those of bacon and feta.

**BROAD BEAN STAGES**

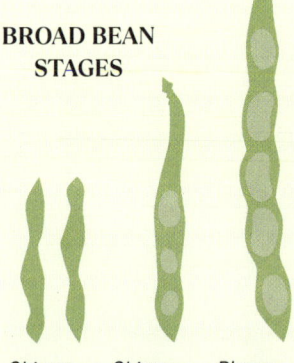

*Skinny like green beans*

*Skinny, do not double-shell*

*Plump, double-shell*

## DOUBLE-SHELLING BROAD BEANS

Remove the beans from the outer pod. Blanch the beans by boiling them for two minutes, then rinse them well in cold water. Pop the tender beans out of their skins.

*I like to add my broad bean shoots to whatever I'm cooking that night. However, if you have a large bowl of them, follow my recipe on page 208 (replacing the snow pea/ mange tout shoots with the same weight of broad bean shoots) to make a tasty stir-fry.*

# Easy broad bean stir-fry

I cook this when I need to use up lots of broad beans in one go. It's best enjoyed when you're in the middle of your bumper broad bean harvest, rather than at the start or end, when the broad beans are a bit more precious.

Double-shell the broad beans (see opposite). Chop the spring onions.

Heat the vegetable oil in a wok over medium heat. Add the broad beans and stir-fry for 3–4 minutes or until the beans turn vibrant green.

Add the water, ensuring that the broad beans are mostly covered. Place a lid on the wok, and simmer for around 5 minutes or until most of the water has been soaked up.

Remove the lid, and add the cooking salt, sugar and chicken bouillon powder. Stir-fry for a further 1–2 minutes. Garnish with spring onion, and serve warm.

**SERVES 2–4**

1.5–2 kg (3¼–4½ lb) broad beans in their pods, to yield 500 g (18 oz) double-shelled broad beans

2 spring onions (scallions), to garnish

1 tablespoon vegetable oil

150 ml (5 fl oz) water

1 teaspoon cooking salt

1 teaspoon sugar

¼ teaspoon chicken bouillon powder

*One of my favourite Chinese condiments is doubanjiang. I use it in my go-to mapo tofu recipe (see page 137). It's a paste made from chillies and fermented broad beans.*

# CELTUCE

*Lactuca sativa* var. *asparagina*
Mandarin: 莴笋 – wō sǔn
asparagus lettuce • celery lettuce • Chinese lettuce • stem lettuce

| | |
|---|---|
| PLANT TYPE | Cool-season annual |
| PLANT FAMILY | Daisy family (Asteraceae) |
| PLANT SIZE | 60 cm (24 in) tall, 30 cm (12 in) wide |
| POT FRIENDLY? | Yes, one plant in a 30 cm (12 in) pot |
| SUN | Full sun or part-sun |
| FROST | Frost tolerant |
| WATER | Regular watering |
| FOOD | Medium feeder |

Celtuce is hands-down one of my favourite Asian greens to grow. An ancient variety of lettuce from southern China, it's cultivated for its long, thick, juicy stem. While the name is a portmanteau of 'celery' and 'lettuce', I find that celtuce tastes more like a blend of broccoli stem and cucumber (which I suppose is like mild celery). It's crunchy, sweet and nutty, and completely delicious.

My parents have been growing celtuce for years because it reminds them of flavours from their homeland. On the rare occasion I see celtuce at Asian grocers (it's only available when it's in season, and is often difficult to find), I think about how special it must feel for immigrants such as my parents to see this and to be able to buy it and bring it home to cook.

Celtuce is hugely popular in China but relatively unknown in many other parts of the world. I'm not sure why celtuce isn't more popular, as it has the potential to be the next big thing. It's easy to grow, extremely versatile and multifunctional, and well-deserving of space in any garden.

## VARIETIES

If you're lucky to have different varieties to choose from, then look out for cold-resistant varieties such as **Purple Sword**, which prefers cooler weather and is great for autumn sowing, and heat-resistant varieties such as **Summer 38**, which is slow to bolt and great for warmer climates and spring sowing. But remember, celtuce tolerates only light frosts.

## GROWING

If you can grow lettuce in your area, then you can grow celtuce. Sow the seeds any time you would sow lettuce seeds. Celtuce needs light to germinate, so sow seeds on the surface of the soil (for direct sowing) or seed-raising mix, press in firmly and water in gently. The seeds need a cool temperature to sprout, so if you're trying to germinate them in late summer, then do this indoors in a cool spot.

Seeds germinate quickly, within a few days. If direct sowing, then thin seedlings out to 30 centimetres (12 inches) apart when they start to look like little lettuces. Enjoy the thinnings as young lettuce. If sowing in punnets, then transplant the seedlings into the garden around 30 centimetres (12 inches) apart once they've grown four adult leaves. In temperate and cool climates, give celtuce a full-sun spot. In warm climates, give it a part-sun spot.

Grow your celtuce in well-drained soil with regular watering, especially on warm days. If you've amended your soil with lots of organic matter before planting, then the plant shouldn't need too much feeding.

When celtuce grows, it starts off like cos lettuce. It's only towards the second half of its life cycle that its stem appears. As the stem grows taller, remove the bottom leaves to eat, and get rid of any leaves that are starting to die (don't worry, this shedding is normal).

Keep an eye out for slugs and snails, as they love celtuce leaves.

*Want to do a bit of interplanting? Celtuce grows quite tall as it matures, and it will have a bare, exposed stem as you harvest the leaves from the bottom.*

*This allows plenty of room for compact plants, such as baby spinach, salad greens, dwarf calendulas and marigolds.*

## HARVESTING

The best part of this plant is its thick stem, which is harvested at the end of the season. However, the leaves can be harvested throughout the growing season and used raw in salads, as well as at the end of the growing season and cooked in stir-fries.

**Leaves** The first opportunity to pick the leaves is when you're thinning your plants. The second chance is when the plants start growing their thick stems. At that time, you can harvest the lower leaves by gently pulling them downwards until they snap off.

**Stems** Celtuce stems are usually ready around two to three months after sowing, and they can grow up to 60 centimetres (24 inches) tall. Harvest the stems when they're at least 30 centimetres (12 inches) tall and 3–4 centimetres (1–1½ inches) thick by cutting them off the base with a hori hori knife (leave the roots behind to break down in the soil). This will ensure that you've got a decent amount of stem once it's peeled. Your last chance to harvest is when you see little flower buds form. After that, the stems will become bitter as they prepare to flower and set seed.

## STORING

Celtuce stems can be stored unpeeled in an airtight container in the fridge. A flat cupcake box is the perfect shape and size for your celtuce stems. If the stems are too long, just cut them in half. Celtuce stems should keep fresh for a week or two. You can also freeze peeled and sliced celtuce stems to use later in cooked dishes.

Celtuce leaves can be stored separately in the same way you'd store lettuce – in a container (with a cloth napkin or paper towel to soak up excess moisture) or in a produce bag in the fridge.

## COOKING

What I love most about celtuce is how quick and easy it is to use. Eat it raw to allow the fresh flavours of the garden to shine, or combine it with other ingredients in a stir-fry. Both the stem and leaves cook quickly, so add celtuce towards the end of the cooking process to tie the whole dish together.

88

### HOW TO PREPARE YOUR CELTUCE STEM

Once the stem has been harvested, remove the bottom leaves that have started to discolour. Compost these, as they won't be good to eat. The rest of the leaves can be set aside for cooking.

To reveal the tasty stem, use a knife to carefully peel off the tough outer skin – it'll slowly come away as you cut. Once you've removed the bulk of it, use a vegetable peeler to remove any remaining white bits (which will be bitter) until only the tender centre is left.

Next, slice off the bottom end. If the centre of the stem is white, then this part of the stem is too mature to eat. Keep cutting off chunks until the centre is no longer white. Once done, you'll be left with a fleshy middle, like a peeled carrot, that is ready to cook.

My favourite way to chop up celtuce is to thinly slice it or cut it into matchsticks, but you can chop it however you like.

## Celtuce-fy your rice

Young celtuce leaves (before the stem matures) can be treated like lettuce. However, older leaves can be a little more bitter and are best cooked. Here's an easy way to use them and it also helps to use up leftover rice – it's an idea from my mum. You can add other fried-rice ingredients to this dish, too, if you like.

Wash your celtuce leaves, and slice them into pieces approximately 1 centimetre (½ inch) wide.

Place the celtuce pieces into a bowl. Massage them with cooking salt, and set aside for 30 minutes. Discard the water that has emerged from the leaves.

Heat the vegetable oil in a wok over medium heat. Add the celtuce, and stir-fry for 2–3 minutes.

Add the rice, and stir-fry until thoroughly reheated. Serve as a fancy side of rice.

**SERVES 2**

Leaves from 2 celtuce plants

1 teaspoon cooking salt

1 tablespoon vegetable oil

3 cups (420 g) leftover cooked rice

## Easy Chinese celtuce salad

This recipe uses raw celtuce. It can be prepared in just 5 minutes (maybe less), and also made ahead of time. Keep the salad in the fridge until you're ready to eat. This allows the flavours to really soak into the celtuce, and the extra chilling time ensures that the salad tastes even more crisp and fresh.

Peel the celtuce stems, and slice them into matchsticks. Chop the spring onion.

Place the celtuce pieces into a bowl. Add the light soy sauce, sesame oil, sea salt and ground white pepper, and mix together.

Garnish with spring onion. Keep refrigerated until you're ready to serve.

**SERVES 2–4**

2 medium celtuce stems

1 spring onion (scallion), to garnish

1 tablespoon light soy sauce

1 tablespoon sesame oil

1 pinch of sea salt

1 pinch of ground white pepper

# CHINESE BROCCOLI

*Brassica oleracea* var. *alboglabra*
Mandarin: 芥兰菜 – jiè lán cài
Chinese kale • gai lan • kai lan

| | |
|---|---|
| **PLANT TYPE** | Cool-season biennial, usually grown as an annual |
| **PLANT FAMILY** | Mustard family (Brassicaceae) |
| **PLANT SIZE** | Around 40 cm (16 in) tall, 30 cm (12 in) wide |
| **POT FRIENDLY?** | Yes, plant one in a 30 cm (12 in) pot or a few in a larger pot |
| **SUN** | Full sun or part-sun |
| **FROST** | Frost tolerant |
| **WATER** | Regular watering |
| **FOOD** | Heavy feeder |

When I was a kid, Chinese broccoli was on high rotation at the dinner table. I always looked forward to a plate of it and loved picking out its young flower buds - the best part of the veggie - because they were just so sweet and tender.

Chinese broccoli looks like a leafy version of broccolini and is enjoyed for its stalks and leaves as well as the young flower buds. The flavour is reminiscent of broccoli, with a little nuttiness and a tiny touch of kale-like bitterness (in a good way). When cooked, both the stalks and leaves retain a deliciously crunchy texture.

The thing I love most about growing Chinese broccoli is how little space it takes up in the garden. Compared to broccoli and broccolini, which are huge plants that grow tall and wide, Chinese broccoli is much more compact (like bok choy). It is also more heat tolerant than other Asian greens and is a great option for warmer climates.

Chinese broccoli and choy sum (see pages 194-7) look very similar, and both are grown for their tender flower stalks. The difference between the two lies in their texture and taste. Choy sum's texture and flavour resemble those of bok choy, whereas Chinese broccoli is similar to broccoli.

## VARIETIES

Choose the variety of Chinese broccoli that is most suited to your climate. If you live in a warmer climate, look for slow-bolting or heat-tolerant varieties.

If you want your Chinese broccoli to grow thick stems, try a late-maturing, long-season variety such as **Kailaan**. Because these varieties take longer to mature, they have more time to produce thicker stems.

If you prefer to harvest Chinese broccoli mainly for its flowers, then select an early-maturing, short-season variety such as **Early Jade** or **Yod Fah.**

## GROWING

Sow seeds on the surface of the growing medium, and cover with a thin layer of vermiculite. Water in gently. The seeds should germinate in about a week. Transplant the seedlings once three or four adult leaves have grown, and plant them 20–30 centimetres (8–12 inches) apart. It takes around seven to nine weeks from seed to harvest.

In warm climates, it's best to grow Chinese broccoli in part-sun to prevent it from bolting early. In temperate and cool climates, give it full sun.

Chinese broccoli plants are heavy feeders, so add nitrogen-rich organic matter to the soil before planting. They also like to be regularly watered. Keep an eye out for caterpillars, slugs and snails.

## HARVESTING

There are a couple of different ways to harvest Chinese broccoli. My personal preference is to wait until the first flower buds appear and shoot up to around the same height as the leaves. Then I harvest the main stem just above a set of leaves, right before the flowers bloom. This triggers the plant to grow side shoots, which can be harvested again in the same way, when new flower buds appear. This method ensures that you have a longer, continual harvest of Chinese broccoli throughout the growing season.

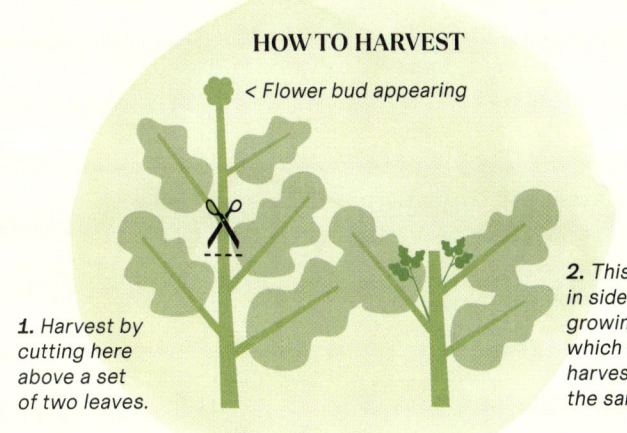

**HOW TO HARVEST**

< Flower bud appearing

**1.** Harvest by cutting here above a set of two leaves.

**2.** This results in side shoots growing, which can be harvested in the same way.

A second way to harvest Chinese broccoli is to remove the entire plant once the first flower buds appear. Compost any fibrous stems; medium-thickness stems can be sliced in half to cook more quickly. This method is simpler but gives you a shorter harvest season.

## STORING

Store Chinese broccoli unwashed in a re-usable container in the fridge for up to a week.

## COOKING

I love to cook Chinese broccoli on its own because it really doesn't need anything else to make it shine. Even in restaurants, Chinese broccoli is served very simply: either in oyster sauce or in garlic. Below, I'll share how to re-create these dishes at home.

## Chinese broccoli in oyster sauce

Cut the Chinese broccoli in half. Finely chop the garlic cloves.

Place the water and ½ tablespoon vegetable oil into a saucepan, and bring to a boil. Add the Chinese broccoli, and bring it back to a boil. Simmer for 1–2 minutes or until tender but not overcooked. Transfer the Chinese broccoli to a plate. I like to stack them on top of each other, facing the same direction.

Discard the water from the saucepan. Heat ½ tablespoon vegetable oil over medium heat. Add the garlic, and cook until fragrant. Add the oyster sauce, light soy sauce and sugar, and mix together. Drizzle the sauce over the Chinese broccoli. Serve warm.

**SERVES 2–3**

300 g (10½ oz) Chinese broccoli

3 garlic cloves

4 cups (1 litre) water

1 tablespoon vegetable oil

1 tablespoon oyster sauce

2 teaspoons light soy sauce

½ teaspoon sugar

## Chinese broccoli in garlic

Cut the Chinese broccoli in half. Finely chop the garlic cloves.

Place the water and ½ tablespoon vegetable oil into a saucepan, and bring to a boil. Add the Chinese broccoli, and bring it back to a boil. Simmer for 1–2 minutes or until tender but not overcooked. Remove the Chinese broccoli from the saucepan and set aside.

Heat ½ tablespoon vegetable oil in a wok over medium heat. Add the garlic, and stir-fry until fragrant. Add the Chinese broccoli, light soy sauce and cooking salt, and mix together. Transfer to a plate, and serve warm.

**SERVES 2–3**

300 g (10½ oz) Chinese broccoli

6 garlic cloves

4 cups (1 litre) water

1 tablespoon vegetable oil

1 teaspoon light soy sauce

¼ teaspoon cooking salt

# CHRYSANTHEMUM GREENS

*Glebionis coronaria* (syn. *Chrysanthemum coronarium*)
Mandarin: 茼蒿菜 – tóng hāo cài
chop suey • crown daisy • edible chrysanthemum • garland chrysanthemum •
shungiku • tung ho

| | PLANT TYPE | Cool-season annual |
|---|---|---|
| | PLANT FAMILY | Daisy family (Asteraceae) |
| | PLANT SIZE | Small, narrow, upright bush to 1 m (3 ft) tall |
| | POT FRIENDLY? | Yes, plant one in a 30 cm (12 In) pot or a few in a larger pot |
| | SUN | Full sun or part-sun |
| | FROST | Frost tolerant once established |
| | WATER | Regular watering |
| | FOOD | Medium feeder |

Whenever winter rolls around, I head to my garden and harvest the edible leaves of chrysanthemum greens. With their distinctive yet mild grassy flavour and tender stems when cooked, they've earned their place as my favourite greens to add to hotpot.

The plant is native to the Mediterranean but eaten as greens across Asia, and it's especially popular in China (where it's called tong hao or tung ho) and Japan (where it's called shungiku). If you've ever grown mums (chrysanthemum flowers), then you'll notice that this plant has similar leaves with deeply serrated edges that stand out in the veggie patch.

Chrysanthemum greens are extremely easy to grow, as they aren't as susceptible to pests and diseases as the greens in the mustard family (Brassicaceae). The plant grows abundantly as a small upright bush, taller than most other Asian leafy greens.

## VARIETIES

There are two varieties of chrysanthemum greens:

1. **Serrate-leaf** (also known as small-leaf, narrow-leaf or medium-leaf) plants have deeply serrated leaves and a stronger flavour than the broad-leaf variety. They are suited to a wide range of climates and tolerant of cooler weather. These plants are most commonly eaten in China and Japan, and they're what I grow.

2. **Broad-leaf** (also known as big-leaf or round-leaf) plants have large, round, broad leaves and a milder flavour than serrate-leaf chrysanthemum greens. As they are tolerant of warm weather, the plants are popular in Southeast Asian cuisines.

## GROWING

Sow the seeds on the surface of the growing medium, and cover with a layer of vermiculite. Water in gently. Baby leaves will pop up within seven to ten days. Once the seedlings have three or four adult leaves, transplant them into the garden about 20–30 centimetres (8–12 inches) apart.

Chrysanthemum greens prefer a mild temperature that is not too cold and not too hot. Compared to other Asian greens, these plants don't mind a bit of shade. My dad always grows his plants in a shady corner of his veggie patch, and they seem to do just fine. Cultivate chrysanthemum greens in a lightly shaded spot in warm climates, as the leaves become bitter when exposed to the hot sun.

Unlike Asian greens found in the mustard family (Brassicaceae), chrysanthemum greens have a bushy, upright growth habit. So, choose their home accordingly, as they can cast shadows over other plants in your garden.

Despite their height, the plants are quite narrow and don't take up too much garden space. I like to grow a few plants together in the corner, so the plant stems keep each other upright. When planted this way, they look like a beautifully abundant mini forest.

## HARVESTING

Chrysanthemum greens grow quite quickly, and you can start harvesting around six weeks after sowing, once plants are around 30 centimetres (12 inches) tall. The young shoots and leaves taste best, so focus on harvesting those.

If you look closely at your plants, you'll notice that they grow from a central stem. To harvest, cut off the growing tips about halfway down the stem, just above where a set of leaves is growing. Harvesting this way encourages more side shoots to grow, which means more greens for future harvests.

All leaves can be eaten, but the larger the leaf, the stronger the flavour. Any tender stems can be eaten, but the more mature ones will be too fibrous and are best composted. Like most cool-season greens, chrysanthemum greens will eventually bolt. If you have the space, let the plants bloom.

### HOW TO HARVEST

*When harvesting, cut about midway down the plant.*

*This encourages side shoots to grow, which results in two main shoots – meaning a bushier plant!*

The flowers are also edible, but you probably wouldn't want to eat the centre (my recent taste tests have found that it isn't the most palatable thing). Instead, use the little petals to garnish salads and decorate cakes.

# STORING

Store chrysanthemum greens unwashed in a re-usable container in the fridge for up to a week. I find that large, flat cupcake boxes are perfect for this, as the stalks can be quite long. If they don't fit, simply cut them in half before storing.

# COOKING

It's best to lightly cook chrysanthemum greens. Overcooking or boiling them for a long time can make them taste bitter.

> **DELIGHTFUL DOUBLE TAKE**
>
> Chrysanthemum greens give you two harvests: **food** in autumn/winter and **cut flowers** in spring. They have the most beautiful yellow and cream daisy-like blooms, and the bees in my garden love them. Cut some of these whimsical little flowers and pop them in a vase for a golden welcome to spring indoors.

## Tung ho tofu salad

Here's an easy cooked salad that's super quick to whip up at home. The recipe uses five spice tofu, which is ready-to-eat tofu that's been marinated in aromatic spices. If you can't find it, use a similarly flavoured firm tofu or simply leave it out.

Finely chop the garlic cloves, and chop the spring onions and chilli (if using).

Place the water and olive oil into a saucepan, and bring to a boil. Add the chrysanthemum greens, and cook for 1–2 minutes over high heat or until the leaves darken in colour (there's no need to bring it to a boil again).

Transfer the chrysanthemum greens to a strainer. Gently apply pressure to drain the excess water. Leave to drain while you prepare the tofu.

Cut the five spice tofu into small squares. Remove the chrysanthemum greens from the strainer, and finely chop.

Place the garlic, tofu and chrysanthemum greens into a large bowl. Add the light soy sauce, sesame oil, sugar, sea salt and ground white pepper. Mix everything together.

Serve the salad warm or cold on a plate, and garnish with spring onion and chilli (if using).

**SERVES 2–4**

2–3 garlic cloves

2 spring onions (scallions), to garnish

1 fresh chilli, to garnish (optional)

4 cups (1 litre) water

1 teaspoon olive oil

300 g (10½ oz) chrysanthemum greens

100 g (3½ oz) five spice tofu

1 teaspoon light soy sauce

1 teaspoon sesame oil

½ teaspoon sugar

¼ teaspoon sea salt

1 pinch of ground white pepper

# Cosy hotpot

Hotpot is our favourite winter dinner tradition at home. It's a do-it-yourself meal that is best enjoyed with friends or family. A large pot filled with delicious soup sits in the middle of the dining table, simmering away on a portable electric stove. Surrounding it are numerous raw ingredients – including my favourite hotpot veggie, chrysanthemum greens, of course – which are cooked one by one over the course of the night and then enjoyed straight away with a special sauce. Hotpot is usually accompanied by long conversations and full bellies, and it's a lovely meal to savour while catching up with loved ones.

If you haven't had hotpot before, then here are a few ideas for your shopping (or sowing) list. Most Asian grocers will have a hotpot section where you can find many of these ingredients:

- **carbs –** instant noodles, glutinous rice cakes, vermicelli, sweet potato noodles
- **meat and eggs –** chicken eggs, quail eggs (we usually just get the tinned ones), meats that have been thinly sliced specifically for hotpot (wagyu beef, regular beef, lamb, pork)
- **mushrooms –** shiitake, shimeji, enoki and oyster mushrooms (be sure to slice all of these into small pieces); halved button mushrooms and sliced portobello mushrooms; I personally love king oyster mushrooms (slice them into thin pieces so they cook quickly)
- **seafood –** crab sticks, sliced fish, fish balls (a classic hotpot ingredient), octopus, prawns, squid, cheese fish tofu (my husband's favourite!), cuttlefish
- **tofu –** classic tofu, firm tofu, silken tofu, tofu skin, bean curd sheets, tofu puffs
- **veggies –** chrysanthemum greens, and pretty much every veggie in this book.

98

**Easy hotpot soup bases** Hotpot soup bases are concentrated, flavoured soup stocks often available at Asian grocers. Unless you're in love with a particular store-bought soup base, why not make your own? It's cheaper, it can help to reduce packaging waste, and there are no preservatives.

On page 101 are my top two homemade hotpot soup bases. I usually make my soup base right before we get our hotpot going, but you can also create a concentrated version and freeze it for later – just reduce the amount of water. My soup bases are lighter in taste than store-bought ones, and will quickly take on the flavours of the ingredients as the night goes on. Each of the soup bases on page 101 will fill one side of a standard split-pot hotpot; you can scale the recipes up or down depending on the type and size of pot you're using.

*Honestly, if you find that it's too much work to create a soup base, then you can just start your hotpot with boiling water. While this will be plain at first, it will quickly become tasty as it soaks up the various flavours of the ingredients being cooked.*

### ZERO WASTE TIP

After a big hotpot session, you'll be left with a rich and tasty soup that is packed with flavour – you've basically been creating an incredibly concentrated soup stock while you eat. You've just applied the permaculture principle of stacking functions! My favourite thing to do with this soup is to turn it into noodle or wonton soup the next day.

You can also reheat it in a saucepan on the stove and have hotpot for one the next day. Pop in whatever leftover hotpot ingredients you have in the fridge, as well as any veggies from your patch, and eat straight from the saucepan while it's simmering on the stove. It's a great way to reduce food waste, even though it's not the way hotpot is traditionally enjoyed. If you haven't got any immediate plans to cook with the soup, then just pour it into small, clean jars, and freeze it for later use.

**OPTION 1**

# Mushroom soup

For the times when I want a base without chilli, or we're using a split-pot hotpot with two soup options, I'll make this mushroom soup. It's vegan-friendly and also great for people who don't like spicy food.

Peel the garlic cloves, but don't chop them. Roughly chop the spring onion or coriander.

Combine all of the ingredients in a pot, and bring to a boil. Simmer for 5 minutes before transferring the pot to your electric stove and beginning your hotpot session.

**MAKES 6 CUPS (1.5 LITRES)**

4 garlic cloves

1 spring onion (scallion) or
	1 sprig of coriander (cilantro)

3–4 shiitake mushrooms

3–4 black wood ears

2 tablespoons light soy sauce

1 tablespoon vegetable oil

6 cups (1.5 litres) water

**OPTION 2**

# Sichuan chilli-style soup

Inspired by a packet of hotpot soup base that I once bought and loved, this base is our go-to for warming us up in winter. It features doubanjiang, a fermented broad bean chilli paste that is available in the marinades section of Asian grocers. You can also use doubanjiang to make mapo tofu (see page 137), and it will last for months in the fridge. The spice level of this soup base is medium – you can add more or fewer spicy ingredients according to taste.

Peel the garlic cloves, but don't chop them. Roughly chop the spring onion or coriander.

Combine all of the ingredients in a pot, and bring to a boil. Simmer for 5 minutes before transferring the pot to your electric stove and beginning your hotpot session.

**MAKES 6 CUPS (1.5 LITRES)**

4 garlic cloves

1 spring onion (scallion) or
	1 sprig of coriander (cilantro)

3 tablespoons doubanjiang

1 tablespoon dried Sichuan
	chilli peppers

1 tablespoon Sichuan
	peppercorns

1 tablespoon vegetable oil

2 star anise (optional)

2 bay leaves (optional)

6 cups (1.5 litres) water

# DAIKON

*Raphanus sativus* var. *longipinnatus*
Mandarin: 白萝卜 – bái luó bo
Chinese radish • Japanese radish • oriental radish • white radish • winter radish

| | | |
|---|---|---|
| | **PLANT TYPE** | Cool-season annual |
| | **PLANT FAMILY** | Mustard family (Brassicaceae) |
| | **PLANT SIZE** | 30–40 cm (12–16 in) tall, 30 cm (12 in) wide |
| | **POT FRIENDLY?** | Yes, plant one in a 30 cm (12 in) pot or a few in a larger pot; choose deep pots |
| | **SUN** | Full sun or part-sun |
| | **FROST** | Frost tolerant (the degree depends on the variety) |
| | **WATER** | Regular watering |
| | **FOOD** | Medium feeder |

Daikon is a group of radishes characterised by large, long, white roots that look like a cross between a turnip and a giant carrot. The name daikon comes from two Japanese words: *dai*, which means 'large', and *kon*, which means 'root'. Compared to red radishes, which are small with a peppery flavour and used in salads, daikon radishes are mild and sweet, and are more commonly cooked and used in soups, stir-fries and pickles.

The best thing about growing daikon is that every part of the plant is edible. When raw, the root is refreshing, crisp and crunchy. When cooked, it becomes translucent, watery and soft, and takes on the flavours of other ingredients. The leaves can be used like spinach, while the seed pods can be eaten raw in salads, pickled or cooked in stir-fries, and have a mild radish flavour.

*White Icicle is not exactly considered a daikon, as it originates from the Mediterranean. But it is a skinny white radish with a similar flavour to daikon. As a bonus, it grows more quickly and is more heat tolerant than daikon.*

## VARIETIES

Larger daikon varieties take longer to grow and are best planted earlier, but they are well worth the wait.

- **Chinese Green Meat Luobo** – is an heirloom Chinese variety. It's a large plant with short roots that have a green skin and flesh, and white tips.
- **Minowase** – is an heirloom Japanese daikon that can grow roots up to 60 centimetres (24 inches) in length – huge!
- **Miyashige** – is another popular Japanese daikon. Its roots grow 30–40 centimetres (12–16 inches) long and have a carrot-like shape.

## GROWING

As daikon is a root veggie, it's best to direct sow the seeds exactly where you want them to grow. Transplanting increases the chance of growing wonky roots. Add lots of organic matter before planting, and keep the soil moist until the seeds germinate. You can expect to see the seedlings pop up within a week or two. Sow seeds every three weeks for a continual harvest.

Space daikon plants 20–30 centimetres (8–12 inches) apart, depending on the mature size of the variety you're growing. While the root itself doesn't take up much space, the leaves on top can spread out quite a bit.

Daikon is often referred to as winter radish because the plants do best when they're given shortening days and cooler temperatures to mature. So, it's best to plant daikon in early autumn and allow it to mature in winter, rather than planting it in spring and allowing it to grow through summer.

Plant daikon in a full-sun position for fastest growth (daikon tastes better when it grows quickly). Loose soil is best if you want your daikon root to grow straight, although daikon is often grown for the purposes of breaking up compacted or heavy clay soil. Keep the plant well-watered, as thirsty plants can become woody.

Daikon plants will start off by growing lots of leaves, so it's normal not to see any roots forming at this stage. It's only towards the end of the growing season that the plants focus on growing roots.

Sometimes, daikon bolts before it has a chance to form a proper root. If this happens, any existing root will be bitter and woody. You can, however, harvest the seed pods and eat them – use them in the same way as daikon root in soups, stir-fries and pickles. Don't worry about saving the seeds from these plants, as you'd end up cultivating plants that were prone to bolting. Instead, try planting a few weeks earlier the following year to give the plants more time to grow.

# HARVESTING

Compared to small, fast-growing red radishes that can be ready to harvest within four to six weeks, daikon takes doubly long to mature; it's ready to harvest in eight to twelve weeks. This is because the roots are bigger, so there's more to grow. As a general rule, smaller daikon varieties mature more quickly than larger daikon varieties.

As daikon matures, the 'shoulder' or top of the radish will start to pop up from the soil. From there, you can see how big it's grown. Harvest your daikon by holding on to the base of the stem and gently pulling it out. If you're growing a big variety and you're worried it'll break, then loosen the soil around it with a garden fork before gently pulling it out. If you're not sure whether your daikon is ready, then err on the side of picking earlier rather than later. Young roots will always be tender, but older roots run the risk of becoming woody.

104

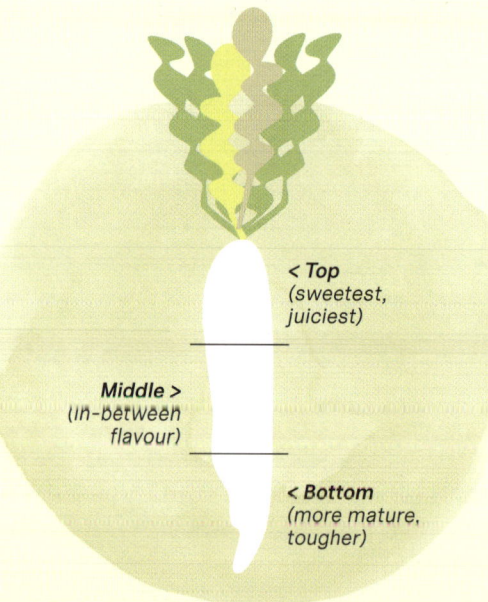

< **Top**
*(sweetest, juiciest)*

**Middle** >
*(in-between flavour)*

< **Bottom**
*(more mature, tougher)*

## ANATOMY OF A DAIKON

- **Top** – This is usually the sweetest, juiciest and most tender part, as it's the youngest part of the daikon. It's best eaten raw or pickled.
- **Middle** – This bit is most versatile and features a mild in-between flavour.
- **Bottom** – This is the more mature part of the radish and can be tougher than the rest. It's ideal for soups and stews.

## STORING

When you harvest your daikon, don't forget that you can eat the leaves as well as the roots. Daikon is best stored with its leaves removed; store the leaves and roots in separate re-usable containers in the fridge to keep them fresh.

## COOKING

It's common for daikon to smell like bad eggs when you're cooking or pickling it. This is because it contains a high number of sulphur compounds. Don't worry, the smell dissipates quickly!

## Braised daikon

I'm a big fan of single-veggie dishes because they're perfect for home gardeners. This daikon dish is a great way to get to know the cooked texture of daikon and enjoy how it takes on the flavours of other ingredients.

Peel the daikon and cut lengthways into quarters, then slice into pieces approximately 1 centimetre (½ inch) wide. Chop the garlic cloves and spring onions.

Heat the vegetable oil in a wok over medium heat. Add the garlic, and stir-fry until fragrant. Add the daikon, and stir-fry for 2 minutes (ensure that it's coated in oil).

Add the light soy sauce, dark soy sauce, chicken bouillon powder, sugar, cooking salt and ground white pepper, and stir to combine. Add the water, then cover the wok with a lid and simmer for 10 minutes.

Remove the lid. Continue to stir and cook for 5–10 minutes or until most of the water has evaporated and the daikon is tender.

Stir through the spring onions and serve.

**SERVES 2-4**

500 g (18 oz) daikon (approx. one medium daikon)

2–3 garlic cloves

2 spring onions (scallions)

1 tablespoon vegetable oil

½ teaspoon light soy sauce

½ teaspoon dark soy sauce

¼ teaspoon chicken bouillon powder

½ teaspoon sugar

½ teaspoon cooking salt

1 pinch of ground white pepper

½ cup (125 ml) water

## Quick namasu

Namasu is a traditional Japanese daikon and carrot salad pickled in sweet vinegar. It's delicious as an appetiser, and I also enjoy mixing a spoonful of it with my rice. If you have furikake (a Japanese rice seasoning made with sesame seeds and seaweed) on hand, then sprinkle it over the rice for an extra-tasty treat.

To store the namasu, you'll need a clean and sterilised glass jar with a lid (see page 72). When you crack open your jar of namasu, you'll notice the bad-egg smell I mentioned earlier – don't worry, it won't linger for long.

Peel the daikon and carrot, and slice both into thin matchsticks.

Place the daikon and carrot into a bowl. Massage them with the sea salt, and set aside for 30 minutes to draw out the water. Drain the water, and transfer the daikon and carrot to the glass jar.

Place the rice vinegar and sugar into a cup. Mix until the sugar has dissolved (heat it a little in the microwave if necessary).

Pour the liquid into the glass jar, and pop on the lid. Before eating, place the jar into the fridge so the namasu can marinate for a day.

**MAKES 500 G (18 OZ)**

300 g (10½ oz) daikon (approx. half a medium daikon)

1 small carrot

1 teaspoon sea salt

2 tablespoons rice vinegar

2 tablespoons sugar

### ZERO WASTE TIP

The leafy tops of the daikon plant – known as radish greens – offer gardeners an exciting culinary bonus. Try these two ideas for cooking your radish greens:

1. **Simple stir-fry** – Cook them in vegetable oil, with light soy sauce and cooking salt to taste, or use radish greens in place of any other leafy greens in the stir-fry recipes in this book.

2. **Radish greens soup** – Boil radish greens, silken tofu and other seasonal veggies in bone broth. Any leftover daikon can be chucked in there as well.

# GARLIC CHIVES

*Allium tuberosum*
Mandarin: 韭菜 – jiǔ cài
Chinese chives • Chinese leek • oriental garlic

| | PLANT TYPE | Perennial |
|---|---|---|
| | PLANT FAMILY | Onion family (Alliaceae) |
| | PLANT SIZE | 30 cm (12 in) tall, 20–30 cm (8–12 in) wide |
| | POT FRIENDLY? | Yes, one clump in a 30 cm (12 in) pot |
| | SUN | Full sun or part-sun |
| | FROST | Frost hardy |
| | WATER | Regular watering |
| | FOOD | Light feeder |

Although garlic chives are technically herbs, they're used more like veggies in Asian cooking. They're perennials related to garlic and spring onion (scallion) that will come back year after year, without much care on your part. Garlic chives are fabulous, low-maintenance plants to grow.

While garlic chives are often grown as ornamentals outside of Asia (they have the most gorgeous little white star-shaped flowers), in Asia they're more commonly grown for their edible leaves, which add a subtle garlic flavour to dishes. The garlicky aroma of the foliage can help to repel caterpillars, slugs and aphids, making garlic chives great companion plants to scatter around your garden.

From afar, garlic chives look like regular chives. The main difference is that regular chives have thin, hollow stems, whereas garlic chives have flat blades that resemble grass – so much so that my mum once harvested garlic chives from my dad's garden and, after she cooked them, my dad suspected that she had accidentally harvested some grass as well! The lesson here is to plant your garlic chives well away from lawn edges so you don't get them mixed up!

# GROWING

The easiest way to grow garlic chives is to divide established plants into smaller clumps. Plants started this way grow more quickly than those started from seed. Garlic chive plants continue to grow larger over time and eventually struggle and lose their vigour, so it's best to divide them every three or four years to keep the plants healthy and productive. It's a great opportunity to introduce a friend to this wonderfully hardy perennial! (If you don't already have garlic chives, then ask a fellow gardener if you can have one of their divided clumps.)

This is how to divide a garlic chive clump:

1. Dig up the garlic chive clump in spring or autumn.
2. Gently tease it apart to separate it into smaller clumps.
3. Replant each small clump 2–3 centimetres (¾–1 inches) deep and 15 centimetres (6 inches) apart. This will turn into an entire carpet of garlic chives in a few years' time.

Garlic chives prefer a full-sun location but will happily live in a part-sun spot as well. While they are drought tolerant, they prefer moist soil and will grow more quickly this way. They're light feeders, so you only need to top up the soil with some organic matter at the start of the growing season.

If you live in a cool climate, don't worry if your garlic chive plants start to lose their leaves. They will die back and become dormant in winter but will start growing again when the weather warms up in spring.

While garlic chives are great at repelling pests (such as aphids) from other plants, they can also, ironically, be susceptible to aphids themselves. Keep an eye out for black aphids – a species that loves onion-family plants – and remove them with a jet of water to prevent them from spreading.

# HARVESTING

When you first plant out a small garlic chive clump, it's best to allow it to establish for a year before you begin to harvest from it. Once it's established, you can start harvesting by cutting leaves at the base. The more you harvest, the more it'll grow.

Towards the end of summer, your garlic chives will start to flower. Unopened flower buds are edible and considered a delicacy throughout Asia. Tender and tasty, the buds are one of my mum's favourite things to cook. If you don't want to eat them, then leave them on the plant for a stunning display of starry white flowers. Pollinators love these blooms and will be *perennially* grateful.

Once the flowers start to die down, it's best to snip them off before they set seed to prevent baby plants from sprouting everywhere. You can save the seeds by shaking them into a paper bag, but it's easier just to divide mature clumps to get more plants.

## STORING

Store garlic chives unwashed in a re-usable container in the fridge, with a cloth napkin to absorb moisture, for up to a week. If you want to preserve your harvest for use over a longer period, then you can freeze it. Wash, dry and chop your garlic chives, then pop them in a thin layer on a baking tray. Place the tray into the freezer. Once the garlic chives are frozen, transfer them to a re-usable freezer bag. Frozen garlic chives store well for up to six months. Alternatively, you can chop and dehydrate the garlic chives, then store them in an airtight jar in the pantry for up to a year – this saves freezer space for other things.

## COOKING

While garlic chives can be eaten raw, they taste best lightly cooked. Always add them at the end of the cooking process, as they lose their flavour when overcooked.

### Garlic chive egg scramble

This is a dish that you'll rarely see in a restaurant, but it's commonly served up in Chinese homes. I grew up eating this a lot with rice, and I love how quick and easy it is to make. It's also delicious!

Chop the garlic chives into 5 centimetre (2 inch) lengths.

Crack the eggs into a large bowl. Add ¼ teaspoon cooking salt, and beat until the yolks and whites are combined.

Heat ½ tablespoon vegetable oil in a wok over medium heat. Pour in the eggs, and cook for 1 minute. Use a spatula to break up the mixture (like you're making scrambled eggs). Once the eggs are lightly cooked (don't overcook them!), transfer them to a plate.

Heat another ½ tablespoon vegetable oil in the wok over medium heat. Add the garlic chives and ¼ teaspoon cooking salt. Cook for 1 minute or until the garlic chives start to wilt. Add the egg back in, mix together and serve.

**SERVES 2-3**

100 g (3½ oz) garlic chives

3 eggs

½ teaspoon cooking salt

1 tablespoon vegetable oil

**ZERO WASTE TIPS**

Extra wonton or dumpling skins can be cut into strips and used in soups, hotpot or noodle dishes. Think of them like thin noodle pieces. Any extra filling can be cooked like minced meat and mixed with rice for a simple fried rice, or used as an extra ingredient in Shanghai cai fan (see page 79).

# Pork and chive dumplings

If you love dumplings, then you've got to try making your own. This is my go-to dumpling recipe and can be used to make wontons as well. The filling can be mixed ahead of time, popped in the fridge to let the flavours meld, then taken out when you're ready to wrap the dumplings.

I like to double this recipe to make a big batch and freeze the extras. If you want to do that, too, then place your dumplings onto an oven tray in the freezer so they don't stick to each other. Then, when frozen, transfer them to a re-usable container or freezer bag. Frozen dumplings can be cooked in the same way as fresh ones; it'll just take a bit longer.

Finely chop the garlic chives.

Place all of the filling ingredients (except the garlic chives) into a large bowl, and mix very well to combine. Add the garlic chives, and mix again. Set the bowl aside for 10 minutes to allow the flavours to meld.

To make the dipping sauce, mix the light soy sauce and black vinegar together in a small bowl. Add the chilli oil, finely chopped ginger and/or sesame oil if desired – the more flavours you add, the better the sauce tastes.

To wrap the dumplings (see pages 112–13), first prepare a small dish of water. Working with one wrapper at a time, place approximately 2 teaspoons of the filling into the centre of a wrapper. Dip a clean finger into the water, then use it to moisten the outside edge of the wrapper. Fold the wrapper in half, then create little pleats as you press it shut. Repeat with the remaining wrappers and filling.

Bring a saucepan of water to a boil. Add the dumplings in batches, cover the saucepan with a lid and bring back to a boil. As the water heats up again, stir the dumplings every now and then so they don't stick to each other or the bottom of the saucepan. Keep an eye on your saucepan – if it overflows, then turn down the heat a little and leave the lid slightly ajar.

Once the water is boiling again, reduce the heat to medium and simmer for 5 minutes or until the dumplings start to float. Let them float for a few minutes, then tear one open to confirm it's cooked through. Remove the dumplings from the water, and serve with the dipping sauce.

**MAKES 40–50 DUMPLINGS**

### For the filling

100 g (3½ oz) garlic chives

300 g (10½ oz) pork mince

2 teaspoons light soy sauce

2 teaspoons vegetable oil

1 teaspoon finely chopped ginger

1 teaspoon oyster sauce

1 teaspoon sesame oil

½ teaspoon cooking salt

¼ teaspoon ground white pepper

2 tablespoons water

### For the dipping sauce

2 tablespoons light soy sauce

2 tablespoons black vinegar

1 tablespoon chilli oil (optional; see page 135)

1 tablespoon finely chopped ginger (optional)

1 teaspoon sesame oil (optional)

### For the wrappers

500 g (18 oz) dumpling/ wonton wrappers (a 500 g/ 18 oz bag usually has around 48 wrappers)

**5**

**6**

# SNOW PEA (MANGE TOUT)

*Lathyrus oleraceus*
Mandarin: 雪兰豆 – xuě lán dòu
Chinese pea • Chinese pea pod • kinusaya

| | | |
|---|---|---|
| **PLANT TYPE** | Cool-season annual | |
| **PLANT FAMILY** | Legume family (Fabaceae) | |
| **PLANT SIZE** | Climbing vine to 2 m (6½ ft) tall; dwarf varieties also available | |
| **POT FRIENDLY?** | Yes, grow multiple plants spaced 10 cm (4 in) apart in a rectangular trough | |
| **SUN** | Full sun or part-sun | |
| **FROST** | Frost hardy | |
| **WATER** | Regular watering | |
| **FOOD** | Light feeder | |

Thought to have originated in southwestern Asia, snow peas are a type of pea with edible pods. The pods are eaten whole and are sweet, juicy and crisp. Snow peas are an Asian favourite in stir-fries but are also delicious steamed or raw in salads. I love to eat them as a refreshing snack while I'm working in the garden.

Snow peas grow on narrow climbing vines, and I love to grow mine up garden arches. Plants can be cultivated very close together and will produce a large number of pods throughout the growing season. The narrowness of the vines makes them an ideal crop for spaces where other plants don't fit. My dad grows his plants in a crack of soil along the fence line!

Being a part of the legume family (Fabaceae), snow peas can forage for their own nitrogen. This means that they are light feeders, so they are both easy to look after in pots and perfect for growing in poor soils.

## VARIETIES

Most snow peas climb up to 2 metres (6½ feet) tall; however, dwarf varieties also exist, and they grow 60–100 centimetres (24–40 inches) tall. Here are some of my favourites:

- **Golden Podded** is a unique variety with yellow pods. It grows vigorously and fruits prolifically.
- **Yakumo Giant** is an heirloom variety from Japan that grows large pods up to 14 centimetres (5½ inches) in length. It's also an early-season variety, so it produces pods more quickly than others.
- **Mammoth Melting** is another large-pod variety. It's more disease resistant than others.

## GROWING

It's best to sow snow peas directly, as their delicate stems can be too fragile to transplant. If you do start them in seed-raising punnets, then transplant the seedlings while they're young and easy to manage.

Sow the seeds around 2 centimetres (¾ inch) deep. Plants can be planted close to each other – leave just 10–15 centimetres (4–6 inches) of space between them. Keep the soil moist but do not overwater, otherwise the seeds can rot. Expect seedlings to pop out of the soil in one to three weeks. For a continual supply of snow peas, sow a new batch of seeds every three to four weeks.

If you live in a temperate or cool climate, then grow snow peas in full sun. If you live in a warm climate, then grow snow peas in part-sun because they require cooler temperatures to flower and set pods.

Taller climbing varieties need something to scramble up, but even dwarf varieties appreciate some support. Initially, you can tie the young plants to a trellis to help the tendrils find their way. Once they do, the plants can be left to climb on their own. Pinch out the growing tips once they reach the top of the trellis to encourage the plants to branch out below.

Snow peas are susceptible to powdery mildew and other fungal diseases. If you have these issues in your garden, then try a disease-resistant variety. Also, it's best not to use pea straw mulch when growing them, as the fungi that break down pea plant material to create pea straw mulch can also spread to snow pea plants.

## HARVESTING

Snow peas can be harvested as soon as they look big enough to eat. Harvest them while the peas inside the pods are still small, and before the pods become large and swollen. The smaller the pods are, the more tender they will be.

## STORING

In times of snow pea abundance, freeze the extras in containers to use later.

## COOKING

Young and tender snow peas are usually stringless. However, older ones may have a fibrous string down the side that will need to be peeled off before cooking.

*The growing tips that you pinch out are edible and make a delicious addition to your next stir-fry. If you love snow pea shoots, then turn to pages 206-9 to see how you can cultivate them in abundance indoors.*

### Snow pea and shrimp stir-fry

This is a super-easy stir-fry that is perfect for weeknights after a busy day at work. If you don't have shrimp, then you can use meat strips or sliced sausage. For vegetarians and vegans, replace the shrimp with 100 grams (3½ ounces) firm tofu or 90 grams (3 ounces) shiitake mushrooms (approximately three mushrooms). Or just use more snow peas (mange touts)!

Top and tail the snow peas. Finely chop the garlic and ginger.

Bring a saucepan of water to a boil. Add the snow peas and cook for 1 minute, then remove the snow peas from the saucepan.

Heat the vegetable oil in a wok over medium heat. Add the garlic and ginger, and stir-fry until fragrant.

Add the shrimp. If you're using raw shrimp, then stir-fry until they start to colour and curl up. If you're using cooked shrimp, then cook until heated through.

Add the snow peas, cooking salt and ground white pepper, and stir-fry for a further 1 minute.

Transfer to a plate, and garnish with sesame seeds.

**SERVES 2-3**

250 g (9 oz) snow peas (mange touts)

2 garlic cloves

1 cm (½ in) piece of ginger

1 tablespoon vegetable oil

100 g (3½ oz) raw or cooked shrimp (approx. 6 shrimp)

¼ teaspoon cooking salt

1 pinch of ground white pepper

Sesame seeds, to garnish

# WOMBOK

*Brassica rapa* subsp. *pekinensis*
Mandarin: 大白菜 – dà bái cài
Chinese cabbage • napa cabbage

| | |
|---|---|
| PLANT TYPE | Cool-season annual |
| PLANT FAMILY | Mustard family (Brassicaceae) |
| PLANT SIZE | 20–30 cm (8–12 in) tall, 15–25 cm (6–10 in) wide |
| POT FRIENDLY? | Yes, one plant in a 30–40 cm (12–16 in) pot |
| SUN | Full sun |
| FROST | Frost tolerant |
| WATER | Regular watering |
| FOOD | Heavy feeder |

A dense, heavy, oval-shaped cabbage, wombok has a mellow flavour quite unlike that of European cabbages. The outer leaves are green, but the inner leaves are creamy yellow and especially sweet. The leaves are great for soaking up flavour, so I add them to my dumpling fillings, stir-fries and stews. The thick, crunchy stems and soft crinkly leaves are suitable for pickling and fermenting. Kimchi is made from wombok, and you can use wombok in suan cai as well (see page 74).

If you like growing cabbages but find that European varieties take up too much space in your urban garden, then consider giving wombok a go. You can grow four womboks in the space of one European cabbage, and they grow more quickly as well.

*If you only want to grow wombok for its baby leaves or as a 'cut and come again' veggie, and you don't mind whether your wombok forms a head or not, then you can plant your wombok at any cool time of the year – autumn or spring.*

## VARIETIES

Larger wombok varieties (such as the ones you see in the supermarket) can be as big as an oval-shaped football and weigh up to 3 kilograms (6½ pounds). However, smaller varieties also exist, and I love to grow them because they're much easier to manage. Here are some easy-to-grow small wombok varieties:

- **Mini Wombok and Mini Napa** – These grow to half the size of traditional, larger womboks, and take up less space, too, making them a great option for small gardens. They also offer a serving and storage size that is more suitable for smaller households.
- **Michihili** – This semi-heading, loose-leaf variety is great for beginner gardeners because you don't have to worry about getting it to form a tight head. It can be harvested as a 'cut and come again' green.

For a more traditional, large, barrel-shaped variety, try **Kyoto**. It's a Japanese heirloom that grows up to 3 kilograms (6½ pounds).

If you live in a warm climate and tend to have trouble with bolting womboks, then look for heat-tolerant, slow-bolt varieties.

## GROWING

The most important thing to get right when growing wondrous heads of wombok is choosing the correct time to plant the seeds. Wombok is a cool-season crop, so we would assume that we can plant them both in autumn going into winter, and in spring going into summer, when the weather is cool and mild (at least for temperate and cool climates).

But with wombok, it's a little different. It's best grown only in warm weather going into cold weather, so the seeds should be sown at the end of summer and allowed to grow through to winter (rather than sowing in spring and growing into summer). This is because wombok requires long, warm days to grow and establish, and short, cool days in order to mature and form tight heads. Early spring sowing is possible in cool climates, but growing wombok at this time of the year is more challenging.

Sow your wombok seeds on the surface of the growing medium, and cover with vermiculite. Water in gently. The seeds will germinate within a week, and you can transplant the seedlings when they have four or five adult leaves. Space the plants 30–40 centimetres (12–16 inches) apart, depending on the mature size of your chosen variety.

Add lots of organic matter to the soil before transplanting your wombok seedlings, and ensure that they're in a full-sun position (especially in cooler climates). Keep the plants regularly watered throughout the growing season. All of these things will help your wombok heads to grow as big and tight as possible, before they bolt when the weather warms up in spring.

**Leaf growth** When your wombok plants first begin to grow, you'll notice that the first few leaves are flat and open. These leaves will eventually become the outer leaves of your wombok heads.

Once the weather cools down, new leaves will grow more upright. These are the leaves that make the cabbage head, although you won't be able to see much of the action, since all of the baby leaves are growing inside the outer leaves!

Some gardeners suggest tying up the outer leaves around the inner leaves at this stage, to protect the inner leaves from sunlight exposure. This is known as blanching (different from blanching when cooking) and is often done to help the inner leaves grow lighter in colour, sweeter and more tender. I've tried this, and it didn't make much difference with my plants – but feel free to test it out yourself.

I've never seen slugs and snails adore a plant more than wombok. It is hands-down their favourite crop in my garden every single year, so be sure to anticipate a slimy brigade. (Some people will grow wombok as a sacrificial crop to protect the other veggies in their garden!) Thankfully, slugs and snails can only reach the outer leaves, which end up getting peeled off anyway.

## HARVESTING

Wombok is ready to harvest at around twelve weeks, depending on the variety. You'll know it's ready when the centre of the head feels firm to the touch. Harvest by cutting at the base of the plant. If there's any snail and slug damage, remove those leaves. Otherwise, store it with as many outer leaves attached as possible to help it last longer. If you're growing wombok as a 'cut and come again' veggie, then harvest the outer leaves whenever you like, leaving the inner leaves to keep growing.

## STORING

A well-stored wombok can live in your fridge for up to two months. In northern China, where winters are dry and cold, wombok is often stored outdoors for the entire season. This storage superpower is helpful for gardeners because it means that wombok doesn't immediately beg for your attention once harvested; it can be kept aside and enjoyed later.

## COOKING

Before cooking wombok, remove any outer leaves that look wilted, especially if the wombok has been stored for some time. If you only want to use a small portion, then simply peel off the required number of leaves. This will help to keep the remainder fresh. If you're using quite a lot of wombok, then just cut the whole head in half lengthways.

*Wombok makes a great addition to dumpling fillings. Turn to page 111 for my dumpling filling recipe, and replace the 100 grams (3½ ounces) of garlic chives with 150 grams (5¼ ounces) of wombok for a tasty twist.*

# Homegrown kimchi

It's such a joyful feeling to nurture a wombok from seed to kimchi. This recipe has a medium spice level thanks to the chilli flakes; add more or less chilli depending on your taste. You'll need to use filtered or unchlorinated water, as straight-up tap water can prevent the good bacteria from performing their magic. To store the kimchi, you'll need one or two clean and sterilised glass jars with lids (see page 72).

Remove the outer leaves of the wombok, reserving a couple of fresh leaves to use as a fermentation weight later. Cut the wombok into quarters lengthways, then chop into 2–3 centimetre (¾–1 inch) pieces.

Place the wombok pieces into a large bowl. Add the sea salt, and massage it in until the wombok starts to soften. Add the water, and use a plate to ensure that everything is submerged. Set it aside on the kitchen bench for 3 hours or ideally overnight.

Transfer the wombok to a colander, and retain the brine. Rinse the wombok under cold water a few times, then squeeze out the excess water (it doesn't have to be completely squeezed out). Leave it in the colander to drain.

Chop the daikon (or daikon and carrot) into matchsticks.

Make the kimchi paste by blending all of the ingredients. Place the drained wombok, daikon (or daikon and carrot) and kimchi paste into a large bowl. If there's leftover paste stuck in your blender, add 1–2 tablespoons of the reserved brine, give it a shake and pour this liquid into the bowl as well. Mix everything together using a pair of tongs.

Transfer to the clean and sterilised jar(s), leaving at least 2–3 centimetres (¾–1 inch) of space at the top. Use a spoon and press the veggies down as much as you can, so they're completely submerged. There should be enough liquid to cover everything; if not, then add an extra tablespoon of the reserved brine, ensuring that everything is covered. Grab the outer wombok leaves that you saved earlier; pop them on top, and press them down into the liquid to keep everything else from floating. Place the lid(s) on the jar(s).

Allow the kimchi to ferment at room temperature, away from direct sunlight. Loosen the jar lid slightly every day to allow excess gas to escape. Start tasting your kimchi from day 3, using a clean spoon. Once it reaches your preferred level of sourness, pop it in the fridge where it will keep fresh for 1–2 months – just be sure to always use a clean spoon when dipping into the jar(s).

**MAKES 8 CUPS (2 LITRES)**

1 kg (2¼ lb) wombok (approx. half a large wombok)

3 tablespoons sea salt

4 cups (1 litre) filtered or unchlorinated water

300 g (10½ oz) daikon (or a mix of daikon and carrot)

**For the kimchi paste**

1 tablespoon finely chopped ginger

1 spring onion (scallion)

½ red onion (or 4 shallots)

8 garlic cloves

2 tablespoons fish sauce (or light soy sauce)

½ nashi pear (tastes best, otherwise red apple also works well)

3 tablespoons Korean-style chilli flakes (gochugaru)

## Stir-fried Shanghai rice cakes (nian gao)

These Chinese rice cakes are commonly enjoyed during the Lunar New Year period as a symbol of growth and a better future. I personally make them at any time of the year, simply because they're delicious. Made from glutinous rice flour, they have a sticky, chewy texture.

Chop the garlic clove, ginger and spring onions. Slice the wombok leaves into 1 centimetre (½ inch) wide strips. Thinly slice the shiitake mushrooms.

Bring a saucepan of water to a boil. While waiting for it to boil, heat 1 tablespoon vegetable oil in a wok over medium heat. Add the garlic and ginger, and stir-fry until fragrant.

Add the mince, and stir-fry for 1–2 minutes or until lightly cooked. Add 1 teaspoon light soy sauce, the dark soy sauce and ¼ teaspoon sugar, and stir-fry for 1 minute or until the mince is cooked. Transfer to a bowl.

Heat another 1 tablespoon of vegetable oil in the wok over medium heat. Add the wombok and shiitake mushrooms, and stir-fry for 3 minutes or until lightly cooked. Remove from the heat.

Once the water has boiled, add the rice cakes to the water and cook for 2–3 minutes or until soft. Transfer the rice cakes to the wok, and place the wok back over medium heat. Add 1 teaspoon light soy sauce, 1 teaspoon sugar, the cooking salt and the ground white pepper.

Stir-fry for 2–3 minutes. Add the mince back in, and mix thoroughly to combine. Garnish with spring onion, and serve hot.

122

**SERVES 3**

1 garlic clove

1 cm (½ in) piece of ginger

2 spring onions (scallions), to garnish

300 g (10½ oz) wombok leaves

120 g (4¼ oz) shiitake mushrooms (approx. four medium mushrooms)

2 tablespoons vegetable oil

200 g (7 oz) pork or chicken mince

2 teaspoons light soy sauce

1 teaspoon dark soy sauce

1¼ teaspoons sugar

500 g (18 oz) oval Chinese rice cakes (usually available frozen)

½ teaspoon cooking salt

1 pinch of ground white pepper

### ZERO WASTE TIP

If you have leftover rice cakes, then throw them into your weekly leftovers hotpot on the stove, or into the next soup or stew you make.

# Other Asian greens

In addition to the Asian greens already profiled in the previous pages, there are many other similar plants that are just as deserving of a place in your veggie patch. The following are all cool-season annuals and planted around the same time as bok choy. You can follow the growing guide for bok choy (see pages 77–8) to cultivate any of these veggies. I've noted their unique characteristics or differences, and special ways to use them. So why not go on a taste-testing adventure and see which ones you like best?

## KOMATSUNA
*Brassica rapa* var. *perviridis*
Mandarin: 小松菜 – xiǎo sōng cài
Japanese mustard spinach • spinach mustard

Komatsuna is a leafy green commonly grown in Japan. The name comes from Komatsugawa in Japan, where it was once extensively grown. It's considered a super food in Japan, as it's rich in many vitamins and contains more than triple the calcium of regular spinach.

Komatsuna tastes like spinach but with a hint of mustard. Young leaves taste more like spinach, while older leaves taste more like mustard. It grows a little larger than bok choy, with dark, glossy, spinach-like leaves. While komatsuna is usually planted during the cool season, it's more heat-tolerant than other Asian greens. It's one of the last greens to bolt in my garden. Harvest it at any stage of growth, and use it as you would spinach.

## MIBUNA
*Brassica rapa* var. *japonica*
Mandarin: 京水菜 – jīng shuǐ cài
mibu • mibu greens

Mibuna is another Japanese green, and it's used in similar ways to mizuna (see opposite). Mibuna plants look like mizuna, but have long, slender, smooth-edged leaves. And if you look closely, the leaves grow as little clusters at the base. Mibuna grows quickly and prolifically but prefers a bit more water than other Asian greens and is more prone to bolting.

124

**ASIAN GREENS MEDLEY**

I love to plant a range of different Asian greens every season, and I harvest most of them as 'cut and come again' greens. Often, my harvest basket is filled with a variety of things – ten bok choy leaves, a handful of mizuna, six of the biggest outer leaves from my tatsoi, and so on. If you ever have a harvest basket like this, then know that you can cook the greens together.

Instead of just having one green in a dish, you can throw in your gardener's selection for a subtle mix of flavours. Cooking with a medley of freshly harvested greens is something that only a home gardener has the pleasure of doing – and it makes life more interesting!

# MIZUNA

*Brassica rapa* var. *nipposinica*
Mandarin: 水菜 – shuǐ cài
Japanese mustard • Japanese potherb • kyona •
spider mustard

Mizuna is a well-loved Japanese green with feathery leaves that resemble a beautiful bouquet. It's a mild Asian version of rocket (arugula), so it's great to grow if you like the idea of rocket but find it too peppery (like me).

Mizuna is one of the quickest and easiest Asian greens I've ever grown, and it's frost hardy, too. The seeds germinate within a few days, and the plants are ready to harvest after about six weeks. You can harvest young mizuna leaves for salads – I always see baby mizuna greens in supermarket salad mixes – or more mature leaves for cooked dishes. Mizuna plants can also be grown indoors as microgreens. There are many varieties available, with my favourite being the striking **Purple**, which is slow to bolt.

# TATSOI

*Brassica rapa* subsp. *narinosa*
Mandarin: 塌棵菜 – tā kē cài
Chinese flat cabbage • rosette bok choy •
spinach mustard • spoon mustard

Originating from China, tatsoi is a low-growing plant with glossy, spoon-shaped leaves. Because it looks like bok choy that's gone through a flower press, the plant has an incredibly beautiful form in the garden. It has a mildly mustard flavour, similar to that of bok choy, but is less watery when cooked. Both green- and purple-leafed varieties exist.

# WARM-SEASON

# VEGGIES

# AMARANTH GREENS

*Amaranthus tricolor*
Mandarin: 苋菜 – xiàn cài
Chinese spinach • edible amaranth • hiyu • vegetable amaranth

| | PLANT TYPE | Warm-season annual |
|---|---|---|
| | PLANT FAMILY | Amaranth family (Amaranthaceae) |
| | PLANT SIZE | Grows up to 60 cm (24 in) tall but is often harvested when it's much smaller |
| | POT FRIENDLY? | Yes, grow one to three plants in one 30 cm (12 in) pot |
| | SUN | Full sun |
| | FROST | Frost sensitive |
| | WATER | Regular watering |
| | FOOD | Medium to heavy feeder |

I'm always searching for leafy greens that will grow through the height of summer, when most other plants quickly give up growing and then bolt. One of my best finds is amaranth greens, which come from the same genus as the amaranth flower (a fluffy bloom used in bouquets) and the grain amaranth. The plant originates from South and Central America but is commonly eaten in Southeast Asia and southern China.

Amaranth's native habitat is the tropics, so naturally it thrives in the heat. It grows quickly and easily, and can be harvested as a 'cut and come again' green. It's often referred to as Chinese spinach, as it's a great summer substitute for spinach (just like sweet potato leaves). But unlike the sweet potato plant, it's compact and easy to cultivate in small spaces and pots.

*Amaranth greens are highly nutritious. They're a fabulous source of protein, as well as vitamin C, iron (they have double the amount of regular spinach), minerals and antioxidants.*

## VARIETIES

Amaranth greens can be green (**Green Leaf**), red (**Red Garnet**) or a mix of green and red (**Bicolour**, **Red Leaf**, **Red Beauty**). The different colours are said to vary slightly in flavour, but all of them taste like spinach to me when cooked. I highly recommend giving the red or green-and-red varieties a go in soup because they have the most vibrant leaf colour that creates a magical beetroot-pink liquid when cooked.

## GROWING

It's easiest to direct sow amaranth seeds, as the seeds are extremely small and the plant grows quickly. Sow seeds when the weather warms up. They require a soil temperature of 20–25 degrees Celsius (68–77 degrees Fahrenheit) to germinate, so in cool climates I suggest sowing in late spring or early summer.

Sprinkle the small seeds on top of moist soil, and gently rake them in. Water in, and keep the soil moist until they germinate. Don't worry if you've sown quite densely. Once seedlings reach 5–10 centimetres (2–4 inches) in height, you can thin them out to 15 centimetres (6 inches) apart and enjoy your first harvest as baby greens.

Because amaranth greens grow quickly, they need lots of water. My plants thrive in self-watering pots. They also prefer lots of organic matter to support their quick growth. Give them a full-sun location, then sit back and watch them do their thing. In general, they're fuss-free plants.

## HARVESTING

Start harvesting your plants once they've grown 30 centimetres (12 inches) tall. For your first harvest, cut off the top growing shoots just above two adult leaves. This will encourage the plant to branch out and grow more leaves. Subsequent harvests can be done in the same way. Or alternatively, cut the entire plant about 5 centimetres (2 inches) from the base, and let it grow back. The more you harvest, the more it'll grow.

Amaranth greens will start to flower from mid-summer. Once this happens, leaf harvest time is over. If you've planted only one variety, then let one plant go to seed so you can save the seeds for the following year. If you've grown multiple varieties, then don't worry about saving the seeds – the varieties may have cross-pollinated.

Once the seeds are almost dry, cut the entire plant near the base and place it upside down in a paper bag. This will allow the seeds to fall out naturally. If you plan to grow amaranth greens in the same spot the following year, then you can just let the seeds dry on the plant and fall to the ground. Amaranth self-seeds easily, so new plants are pretty much guaranteed to pop up in the same spot during the next growing season.

## STORING

Store amaranth greens unwashed in a re-usable container in the fridge for up to a week.

## COOKING

Remove the young shoots and leaves from the stems, and use them for cooking. Younger leaves taste best, whereas older leaves have a stronger flavour. While all parts of the greens are edible, the thicker stems can be tough and stringy and should be peeled before eating. Do this by snapping the stems in half, and then pulling off any stringy bits that you see.

Younger amaranth leaves can be enjoyed fresh in salads, but do keep in mind that they (like spinach) are high in **oxalic acid** (too much oxalic acid can increase the risk of kidney stones in some people). Cooking them reduces the oxalate level.

### Pink magic amaranth

I almost always cook amaranth on its own and love using bicolour or red amaranth for the thrill of pink-hued soup. This is a simple recipe for a basket of amaranth greens, which will cook quickly – perfect for summer, when you don't want the stove to heat up the house. You might think that 400 grams (14 ounces) is a big bundle of leaves, but gosh do they shrink like spinach!

Finely chop the garlic cloves.

Heat the vegetable oil in a wok over medium heat. Add the garlic, and stir-fry until fragrant.

Add the amaranth leaves and shoots, and stir-fry for 1–2 minutes or until they have wilted.

Add the light soy sauce, cooking salt and water. Stir to combine, then cover the wok with a lid. Simmer for 2–3 minutes or until most of the water has evaporated. Garnish with sesame seeds if you like, and serve warm.

**SERVES 2–4**

3–4 garlic cloves

1 tablespoon vegetable oil

400 g (14 oz) amaranth leaves and young shoots

1 teaspoon light soy sauce

1 teaspoon cooking salt

100 ml (3½ fl oz) water

Sesame seeds, to garnish (optional)

# CHILLI

*Capsicum annuum*
Mandarin: 辣椒 – là jiāo
hot pepper

| | | |
|---|---|---|
| | **PLANT TYPE** | Perennial; grown as an annual in cool climates |
| | **PLANT FAMILY** | Nightshade family (Solanaceae) |
| | **PLANT SIZE** | Compact bush, up to 70 cm (28 in) tall and 40 cm (16 in) wide |
| | **POT FRIENDLY?** | Yes, and recommended; use a 40 cm (16 in) pot |
| | **SUN** | Full sun |
| | **FROST** | Frost sensitive |
| | **WATER** | Regular watering |
| | **FOOD** | Heavy feeder |

Chilli is a prominent ingredient in many Asian cuisines. People in places such as Southeast Asia, Korea, and Sichuan and Hunan provinces in China (just to name a few) love to add a bit of heat to their dishes. Some of my favourite chilli dishes have been the ones I've devoured during my travels in Asia - from the mouth-watering chilli crab of Singapore to the tangy curries of Thailand.

If you love chilli, then you've got to grow your own. There's nothing better than the vibrant heat of a freshly picked chilli. The plants are compact and happily grow in pots, so are highly suited to small spaces. Just one plant produces dozens of chillies, which can be frozen and used throughout the year. There are so many varieties to choose from compared to the two options I always see in supermarkets - discovering your favourite variety is half the fun.

*Both capsicum (sweet pepper) and chilli (hot pepper) are the same species, just with differing heat levels. If you're also interested in cultivating capsicums, then follow this guide because they're grown in exactly the same way.*

# VARIETIES

The world of chillies is vast, and some 3000+ varieties exist. Every country has its favourite native chillies. Here are few to get you started:

- **Er Jing Tiao** – The most popular Sichuan chilli, it's used to make chilli oil and doubanjiang, and it's often sold dried. Er Jing Tiao chillies are up to 15 centimetres (6 inches) long with a slight curve, and have a moderate heat. This is my go-to Sichuan chilli, and I love using it to make mapo tofu (see page 137) and hotpot soup bases (see page 101).
- **Chao Tian Jiao (Zhi Tian Jiao)** – The name translates to 'facing heaven' or 'facing sky', as the chilli fruits grow pointing upwards. Another popular and prolific Sichuan chilli, the fruit grows 3–6 centimetres (1–2 inches) long but is hotter than Er Jing Tiao.
- **Bird's Eye Chilli** – This fabulous Thai variety is more cold tolerant and ripens earlier than other chillies. I find it a great all-rounder to use in stir-fries and curries. This one's spicier than both Sichuan varieties. The chillies are 2–4 centimetres (¾–1½ inches) in length and grow facing upwards.

# GROWING

Chilli plants require a long, warm growing season, so it's best to sow seeds indoors using a heat mat. You can expect seedlings to pop up within three weeks. Transplant the seedlings into the garden after your last frost. Slugs and snails adore chilli seedlings, so keep an eye out for these pests and remove them before they eat your seedlings to the ground.

You'll only need one plant per variety – unless, of course, you love eating lots of chillies! Varieties that produce larger chillies will grow fewer chillies per plant; varieties that produce smaller chillies will grow more chillies per plant.

Chilli plants grow slowly in cold weather and quickly in warm weather. Give plants a full-sun spot – afternoon sun is especially appreciated by these heat-loving plants. An easy way to remember this is: hot peppers need hot weather. In cooler climates, grow plants near a brick wall that receives afternoon sunlight to create a warmer microclimate.

I always grow my chilli plants in pots. They're happy to be confined in containers, and the portability of pots is handy for overwintering them (see page 134). A 40 centimetre (16 inch) pot is best, but a 30 centimetre (12 inch) pot also works – you'll just get fewer fruits. It's helpful to use a 1 metre (3 foot) stake to keep each plant upright, especially once it starts to bear fruits.

## HARVESTING

Chillies begin to fruit in late summer and can be harvested at any stage of colour. Most chillies are green or yellow to begin with, then turn red as they mature. Some chillies will turn black/purple before they turn red – this is simply part of the ripening process. The redder the fruit, the hotter it is. Harvest your chillies by cutting them off with a bit of stem attached. This helps to keep them fresh for longer.

**Overwintering chilli plants** In warm and tropical climates, chilli plants are grown as perennials and can be harvested all year round. Depending on the variety, they can live for up to 15 years. In temperate and cool climates, chilli plants die when hit by cold weather. However, if you overwinter your plants, then you can grow them as perennials.

Overwintering refers to the practice of protecting sensitive plants from cold weather, especially frost, during winter. If you successfully overwinter chilli plants, then you don't have to sow new seeds every year. You'll have a head start each season!

Start by growing chillies in pots. When winter rolls around and the weather cools down, move the pots to a sheltered area (such as under a carport, in a greenhouse, near a sunny brick wall that retains heat or under the eaves of a house – even indoors).

Chilli plants will become dormant as the weather cools down, and they may get a bit scraggly during winter. Let them be until the weather warms up again, then prune off any dead bits and cut plants down to a few main stems. New shoots will quickly pop out as the plants wake up.

## STORING

The easiest way to store chillies is to freeze them whole. Chillies will keep for up to a year in a container in the freezer, until your next season's harvest. As chillies are quite thin, they don't need to be thawed before use. Just slice and add them straight into whatever you're cooking.

You can also dehydrate chillies. Select only your most perfect, unblemished fruits, as this reduces the chance that they'll go mouldy. Slice large chillies in half lengthways, or chop them into 1–2 centimetre (½–¾ inch) pieces. Smaller chillies can be dried whole. Spread them in a single layer on a tray and use a dehydrator or oven on the lowest temperature. Turn every few hours until they are completely dry and easily crumble between your fingers. Store in an airtight jar in your pantry for up to a year.

134

*Top: Chao Tian Jiao ('facing heaven') chilli plant*
*Bottom: A large chilli harvest like this can be dried and made into chilli flakes or powder*

# COOKING

The general rule with chillies is the smaller it is, the hotter it is. Take care when preparing chillies, especially super-hot ones. Wear gloves, and don't rub your eyes. If you accidentally get chilli in your eye, then milk is a great antidote for the pain. Don't ask me how I know! If you'd prefer to cook with a chilli that isn't too hot, then remove the seeds (where most of the heat is contained) and just use the flesh.

*Despite the spicy heat packed into some chilli varieties, you'll find that wildlife still loves to nibble on the plants - so do keep an eye out.*

## Homemade chilli oil

Chilli oil is a hugely popular Chinese condiment that can be produced with minimal effort at home. The beauty of making your own is that you can experiment with different flavours. I've listed some recommended spices, but feel free to get creative! You can use your homemade chilli oil as a dipping sauce for dumplings and wontons, or mix it into fried rice or noodles.

It's easier to make chilli oil with dried chillies than fresh, so you'll need to dehydrate your harvest first. This recipe has a medium spice level; add more or fewer chillies depending on your taste. To store the chilli oil, you'll need a small clean and sterilised glass jar with a lid (see page 72).

Slice the chillies into small pieces, and dehydrate using the instructions in the Storing section on the opposite page. Once dried, break the chilli pieces into flakes using a mortar and pestle or spice grinder, and place them into a ceramic mug or bowl. Slice the garlic cloves.

Heat the vegetable oil in a saucepan over low heat until it slowly starts to bubble. Add the garlic and your choice of spices (if using), and allow them to infuse for 5–7 minutes while the oil continues to slowly bubble away. Remove from the heat once the garlic has started to brown.

Allow the oil to cool for a few seconds, then slowly pour it through a strainer and onto the chilli flakes, removing the garlic and spices in the process. Take care, as the chilli flakes will start to sizzle and pop.

Gently stir to mix everything together, and set aside to cool.

Once the chilli oil is cool, pour it into the jar. Place the lid on the jar, and allow the chillies to infuse the oil in the fridge for 2–3 days. Use within 3–4 weeks.

**MAKES 130 ML (4½ FL OZ)**

200 g (7 oz) fresh chillies

3 garlic cloves

½ cup (125 ml) vegetable oil (or other neutral oil with a high smoke point)

**Spices for flavouring (optional)**

1 cinnamon stick

1 star anise

1 bay leaf

2 teaspoons Sichuan peppercorns

2 teaspoons sesame seeds

*The reason we pour the hot oil onto the chilli flakes, rather than adding the chilli flakes to the saucepan over the heat, is to prevent the chilli flakes from burning, which changes their flavour.*

# Mapo tofu

A signature dish of Sichuan in China is mapo tofu. It's a rich and spicy dish that's all about the chilli. Add as many or as few chillies as you can handle, and enjoy with rice and veggies.

Slice the tofu into small pieces. Chop the spring onion. Coarsely chop the Sichuan chillies.

In a small bowl, prepare a slurry by mixing the cornflour with the water.

In a large bowl, mix together the marinade ingredients. Add the pork mince and mix. Allow the pork to stand for 10 minutes.

Heat the vegetable oil in a wok over medium heat. Add the Sichuan peppercorns, infuse them in the oil for 1 minute, then remove the peppercorns with a strainer and discard.

Add the pork mince, and stir-fry for 5 minutes or until lightly cooked through. Break the mince into small pieces while cooking.

Add the tofu, and stir in. Add the doubanjiang, light soy sauce, dark soy sauce, sugar, slurry and Sichuan chillies, and stir-fry for 5 minutes. Garnish with spring onion before serving.

**SERVES 4**

500 g (18 oz) tofu (a soft, classic firmness is best)

1 spring onion (scallion), to garnish

Sichuan chillies, to taste

2 teaspoons cornflour (cornstarch)

1 tablespoon water

200 g (7 oz) pork mince

1 tablespoon vegetable oil

1 teaspoon Sichuan peppercorns

2 tablespoons doubanjiang

1 teaspoon light soy sauce

2 teaspoons dark soy sauce

1 teaspoon sugar

**For the marinade**

1 teaspoon light soy sauce

1 teaspoon dark soy sauce

1 teaspoon cooking wine

1 teaspoon finely chopped ginger (optional)

2 teaspoons cornflour (cornstarch)

2 tablespoons water

# CHOKO

*Sechium edule*
Mandarin: 佛手瓜 – fó shǒu guā
chayote · mango squash · mirliton · vegetable pear

| | |
|---|---|
| PLANT TYPE | Perennial |
| PLANT FAMILY | Cucurbit family (Cucurbitaceae) |
| PLANT SIZE | More than 10 m (33 ft) long, spreading vines; needs vertical support |
| POT FRIENDLY? | Yes, one plant in a 40–50 cm (16–20 in) pot, although best grown in the ground |
| SUN | Full sun |
| FROST | Frost sensitive |
| WATER | Regular watering |
| FOOD | Heavy feeder |

One of the easiest climbing vines you can grow is choko, a subtropical fruit that is eaten like a veggie. It originates from Central America but is also enjoyed across Asia. Chokos grow quickly and provide an abundance of fruits in autumn, when most other fruiting veggies have finished up for the season. A single choko plant can grow up to 100 fruits, so you'll never be short of something to eat!

Choko is mild and sweet, and tastes like apple crossed with cucumber. The texture is refreshingly crisp and quite similar to celtuce. It's extremely versatile, and can be used raw in salads, cooked in stir-fries, roasted and even pickled. Every part of the choko plant is edible, from its shoots, leaves and fruits to its tuberous roots.

*Choko plants can be grown in pots, but they suit only large containers that are at least 40 centimetres (16 inches) in diameter. Container-grown chokos will produce fruits but less prolifically than those cultivated in the ground.*

## VARIETIES

There are two main types of chokos: **green chokos** and **white chokos**. Green chokos are more common, while white chokos are much harder to come by. Both taste the same, so it's just a matter of looks.

## GROWING

To start a choko plant, you'll need to get your hands on a choko fruit rather than choko seeds. This is because the seed of a choko germinates while it's still inside the fruit (this is known as viviparous germination).

Choko gardeners will often have plenty of chokos – potentially even sprouting ones – to share. Otherwise, try an Asian grocer or farmer's market – pick a large, mature one for best results. Leave your choko on the bench indoors, and within a few weeks it will start to sprout. Once it sprouts, bury two-thirds of it in a pot of moist soil, with the sprouting end on top. The choko fruit contains enough water and nutrients to sustain the early stages of growth, so don't overwater the plant or it'll rot. As the sprout starts to grow, the fruit below will naturally shrivel away.

Keep your potted choko in a warm, sheltered spot, and plant it out in spring once overnight temperatures are at least 10 degrees Celsius (50 degrees Fahrenheit). Transplant it to a spot with full sun and lots of vertical climbing space – the more, the better.

Choko plants grow luscious green vines with broad leaves during the warmer months of the year, making them a great source of shade when you need it most. Consider planting a choko so it can grow over a pergola or an enclosed chicken run.

If you live in a warm or tropical climate, chokos will be evergreen. In a temperate climate, they will become deciduous in winter, dying back to the roots and re-emerging in spring. If you live in a cool climate with heavy frosts, you'll need to grow chokos as annuals. Growing chokos in temperate/cool climates is actually a good thing, as a shorter season means fewer chokos. You'll appreciate them more and will be less likely to get sick of them.

**Pruning** Choko vines can grow more than 10 metres (33 feet) long but should be pruned to keep the plants productive. In smaller spaces and pots, keep only a few main vines growing from the base, and chop off the tips once they reach the height of the trellis. This will encourage the vines to grow more side branches, increasing yields. Save any young shoots and leaves that you prune off, as they make a fabulous stir-fry.

## HARVESTING

Choko plants grow both male and female flowers, with female flowers having a small, round bulge at the stem that turns into the fruit once pollinated by bees. This fruit takes about a month to grow, and first harvests usually begin around autumn.

As fun as it is to grow huge fruits, smaller chokos are more tender and taste better. Pick chokos when they are anywhere from egg size to tennis ball size. Chokos this size don't need to be peeled before eating and are great for using raw.

Larger chokos can also be eaten and are best peeled and cooked. The tubers can be dug out of the ground and eaten like potatoes.

## STORING

Fresh chokos should be stored in an airtight container or bag, as they need humidity to prevent them from shrivelling. Store them in a cool place indoors. They'll keep for up to eight weeks before they naturally start to sprout.

If you want to store your chokos for longer, thinly slice and dehydrate them like chips. Alternatively, you can slice, blanch and then freeze them.

## COOKING

I usually reserve large chokos for cooking, and use small ones raw. Large chokos are best peeled before cooking. Use a veggie peeler but take care – chokos release a sticky liquid when peeled, making it a bit fiddly. Gloves are a good idea. Slice chokos thinly so they cook quickly, and compost the central core. The seeds of young chokos can be sautéed and have a nutty flavour.

### NOT-A-CHOKO

There's a cheeky choko doppelganger out there that is a toxic, invasive weed. You might encounter it one day and think you've lucked out and found a new foraging hotspot. But before you pick and eat the fruits, be sure to identify the plant properly, as it could be an extremely common look-alike known as the **moth vine** (*Araujia sericifera*). Here's how to tell the two plants apart:

**Choko**

- fruit is firm and heavy, up to 500 grams (18 ounces) for mature fruit
- fruit hangs like a pear (pointed end on the top; larger part on the bottom)
- has large, pumpkin (squash)-like leaves
- Is edible

**Moth vine**

- fruit is small and soft
- fruit hangs like an upside-down pear (larger part on the top; pointed end on the bottom)
- has small, triangular leaves
- is NOT edible

*While chokos are perennials, they only live up to eight years and fruit most prolifically aged three to five. So, it's best to replant them every three or four years to ensure that you always have productive plants.*

*An easy way to cook chokos is to make chips. Peel and slice your chokos into thin wedges. Cover the wedges with flour or tapioca starch, then dip in egg (add your favourite herbs and spices to the egg). Coat with panko or breadcrumbs, and bake at 180 degrees Celsius (350 degrees Fahrenheit) for 20 minutes, turning halfway through cooking.*

# Chicken and choko stir-fry

Chokos make a great stir-fry ingredient and will take up the flavour of whatever you're cooking them with. The key to delicious stir-fried choko is to slice them thinly so they cook quickly and have the right amount of crunch. For a plant-based version, replace the chicken with firm tofu.

Chop the spring onion or coriander.

Thinly slice the chicken, and place it into a bowl. Peel the choko, and cut into quarters. Cut off the white centre and seed from each quarter, then thinly slice the quarters.

In a bowl, mix together the marinade ingredients. Pour the marinade over the chicken, and massage well for 1 minute or until the flavours have soaked in and the chicken has softened. Set aside to marinate for 10 minutes.

Heat 1 tablespoon vegetable oil in a wok over medium heat. Add the chicken, and stir-fry for 2–3 minutes or until lightly cooked through. Set the chicken aside in a clean bowl.

Heat another 1 tablespoon vegetable oil in the wok over medium heat. Add the choko, and stir-fry for 1 minute.

Add the water, light soy sauce, chicken bouillon powder and cooking salt, and continue to stir-fry for 5 minutes or until the choko pieces have softened but are still crisp.

Return the chicken to the wok, and mix it in. Transfer to a plate, and garnish with the spring onion or coriander before serving.

**SERVES 2–4**

100 g (3½ oz) chicken thigh fillet

350 g (12¼ oz) choko (1 whole choko)

2 tablespoons vegetable oil

1 tablespoon water

¼ teaspoon light soy sauce

¼ teaspoon chicken bouillon powder

¼ teaspoon cooking salt

2 teaspoons spring onion (scallion) or coriander (cilantro), to garnish

**For the marinade**

½ teaspoon Chinese cooking wine

½ teaspoon cornflour (cornstarch)

¼ teaspoon light soy sauce

¼ teaspoon garlic powder

¼ teaspoon finely chopped ginger

¼ teaspoon sugar

¼ teaspoon cooking salt

1 pinch of ground white pepper

1 tablespoon water

## OTHER STIR-FRY IDEAS

Choko has a crisp texture and tastes best when combined with other ingredients that have a similar texture or hold their shape well:

- choko + egg
- choko + capsicum (sweet pepper) + black wood ears
- choko + pork mince/chicken/tofu + black wood ears.

Young choko shoots, including the tendrils, are also edible. Here are a few combos to get you started:

- choko shoots + shiitake mushrooms + sesame oil + light soy sauce
- choko shoots + garlic + cooking salt
- choko shoots + egg.

# EGGPLANT (AUBERGINE)

*Solanum melongena*
Mandarin: 茄子 – qié zi
baigan • brinjal

| | | |
|---|---|---|
| 🏷️ | **PLANT TYPE** | Perennial; grown as an annual in cool climates |
| ◊ | **PLANT FAMILY** | Nightshade family (Solanaceae) |
| ⤢ | **PLANT SIZE** | Small bushes around 60 × 60 cm (24 × 24 in), depending on variety |
| ▽ | **POT FRIENDLY?** | Yes, plant one in a 40 cm (16 in) pot |
| ✳️ | **SUN** | Full sun |
| ❄️ | **FROST** | Frost sensitive |
| ♦ | **WATER** | Regular watering |
| ✥ | **FOOD** | Heavy feeder |

Native to subtropical areas of Southeast Asia, eggplant is a delightfully ornamental veggie that suits small-space gardening. It comes in a rainbow of colours – white, green, red, yellow, lilac – and there are stripy and mixed coloured ones, too. Some are large and teardrop shaped, while others are small and round. I always love browsing eggplant seeds, as there are so many varieties available.

In warm climates, eggplant is a short-lived perennial; however, in cool climates, it's grown as an annual (unless you overwinter the plant in the same way as chilli plants; see page 134).

*Eggplant (and chilli) seedlings are quite slow growing, so they're the first warm-season plants I sow.*

# VARIETIES

I personally prefer growing smaller eggplants, as it's easier to use the whole fruit while cooking. These plants produce lots of fruits, whereas larger varieties grow fewer but larger fruits. Here are some popular Asian varieties to check out:

- **Mitoyo** – This variety from Japan has large black fruits.
- **Ping Tung Long** – This is a very popular lilac-coloured Asian eggplant that has long, slim fruits.
- **Thai Purple Ball** – A popular variety in Thailand and Vietnam, it's similar to Thai Round Green but with dark purple fruits.
- **Thai Round Green** – This 5 centimetre (2 inch) round eggplant is green in colour and popular in Thailand.

# GROWING

Like chillies, eggplants require a long, warm growing season. In cool climates, they're best started indoors to get a head start. Sow seeds on the surface of the growing medium, and cover with vermiculite. Use a heat mat to help with germination. The seedlings will appear in one to two weeks. As they're sensitive to frost and cold weather, it's best to transplant seedlings outdoors once the temperature has warmed up a bit and they've developed at least five adult leaves.

Grow the plants in a full-sun position, and they'll thrive. I find that my eggplants especially love the warm afternoon sun. Plants grown in shade or during cool weather will grow slowly and may produce few or no fruits.

Eggplants grow as compact bushes and need to be spaced 50 centimetres (20 inches) apart in the garden. They also make great potted plants, just like chillies (see pages 132–7). It's best to pop in a stake and tie the main branches to it for support, as they can get heavy when they bear fruits.

These plants are heavy feeders, so dig in lots of organic matter before planting and top up a couple of times throughout the growing season, especially when the plants start to flower. They also require regular watering, so be sure to keep your plants hydrated. Thirsty plants will drop their flowers and young fruits.

# HARVESTING

Eggplants are ready to pick when they have glossy skin that is firm to touch, with only a little resistance when you squeeze it. Overly mature fruits will have a dull colour and can be wrinkly and squishy. If you're not sure, harvest one fruit and cut it open. If the seeds inside are still immature, then you've picked it at the right time. If the seeds inside have started to form and turn brown, then you've picked it a little late.

Older eggplants, while edible, are more bitter. You can remove the skin and seeds, and soak the flesh in salty water to draw out some of the bitterness.

Harvest using secateurs, and keep a bit of the stem attached to help the fruit stay fresh. Take care when harvesting, as stems have sharp thorns like roses do!

**Overwintering** Eggplant plants can be overwintered in the same way as chilli plants (see page 134).

# STORING

Eggplants are best used fresh, as they have only a two- to four-day storage life once picked. (If you want to store them for longer, then it's best to preserve them by dehydrating or bottling them.) Store fresh eggplants in a paper bag in a cool spot in the house, out of direct sunlight. Their preferred storage temperature is 10–12 degrees Celsius (50–54 degrees Fahrenheit), but that can be difficult because they're a summer crop! If you don't have a cool spot for them, then it's best just to pop them in the fridge.

# COOKING

If I have a big harvest of eggplants, then I'll steam them because I can use quite a few fruits in one go. If I only have a couple, then I'll add them to stir-fries. One of my favourite stir-fry combinations features eggplants, potatoes, carrots and capsicums (sweet peppers), which boasts a beautiful rainbow of colours. I like to season this stir-fry with a dash of light and dark soy sauce, white vinegar and sugar.

*Eggplant fruits quickly oxidise and turn brown once cut open. This doesn't look very appealing when they're served. So, slice eggplants right before you use them. A simple tip I learned from my mum is to transfer cut pieces straight into a bowl of warm water as you slice them. This slows the oxidisation process and helps to reduce the bitterness.*

# Steamed eggplant with spicy sauce drizzle

This is an extremely simple dish that's packed full of flavour. It's great to cook during peak eggplant (aubergine) season in the garden.

Cut off the ends of the eggplant, and then slice the eggplant into long halves or quarters. Finely chop the spring onion and garlic cloves.

Place the eggplant into a steamer, and cook for 15 minutes or until tender.

While the eggplant is steaming, mix the sauce ingredients together in a small bowl.

Once the eggplant is ready, drain any excess water. Transfer the eggplant to a plate, drizzle the sauce on top and garnish with spring onion. Serve hot.

**SERVES 3–4**

400 g (14 oz) eggplant (aubergine) (a slim variety works best)

1 spring onion (scallion), to garnish

**For the sauce**

2 garlic cloves

2 teaspoons vegetable oil

2 teaspoons sesame oil

2 teaspoons white or black vinegar

1½ tablespoons light soy sauce

1 teaspoon chilli oil (see page 135)

1 teaspoon finely chopped ginger

1 teaspoon sugar

# FLAT BEAN

*Phaseolus vulgaris*
Mandarin: 扁芸豆 – biǎn yún dòu
helda bean • romano bean

| | | |
|---|---|---|
| 🪴 | **PLANT TYPE** | Warm-season annual |
| 🌢 | **PLANT FAMILY** | Legume family (Fabaceae) |
| ⤢ | **PLANT SIZE** | Climbing varieties: up to 2 m (7 ft) tall; bush/dwarf varieties: up to 40 cm (16 in) tall |
| 🪣 | **POT FRIENDLY?** | Yes, climbers in a large container with trellis; bush/dwarf varieties in a 40 cm (16 in) pot |
| ✳ | **SUN** | Full sun |
| ❄ | **FROST** | Frost sensitive |
| 💧 | **WATER** | Regular watering |
| ⁙ | **FOOD** | Light feeder |

The flat bean is common in Europe and also popular in Asian cooking. The pod is long, wide and flat, with aesthetically pleasing wavy edges. It looks like a cross between a long bean and a snow pea (mange tout).

Flat beans are eaten whole - pods and all - and most varieties are stringless. They have a fresh and crisp texture and make a great summer substitute for snow peas.

Compared to long beans, which need a long, hot summer to grow, flat beans will grow anywhere other common green beans (such as French beans and shelling beans) grow. This makes them easier to cultivate than long beans through a cool-climate summer.

*I like to sow a second batch of flat bean seeds approximately one month after the first. This helps to extend the harvest period through summer.*

# VARIETIES

Both climbing and bush/dwarf varieties are available. Climbing plants tend to produce beans over a long season, whereas bush/dwarf plants tend to produce their beans over just two or three weeks. Bush/dwarf varieties are great for when you don't have vertical space, but they produce fewer beans per plant than climbing varieties.

For a true Asian flat bean, look for Chinese heirloom varieties (such as **Qing Bian** or **Lu Qing**). Otherwise, European varieties are just as good. **Giant of Stuttgart** is a climbing variety that grows pods up to 30 centimetres (12 inches) long.

# GROWING

Like most beans, flat beans usually do best if direct sown, as they don't transplant well. Because flat beans don't require a long growing season, it's not essential to start them indoors (unless you live in a really cold climate).

Flat beans can be grown quite close together. Plant the seeds of climbing varieties approximately 15 centimetres (6 inches) apart, and the seeds of bush/dwarf varieties approximately 30 centimetres (12 inches) apart. The seeds will naturally start to germinate once the soil temperature is at least 16 degrees Celsius (61 degrees Fahrenheit). From then, the seeds will take around five to ten days to sprout.

Plant flat beans in a sunny spot. If you've sown a climbing variety, give it something to scramble up (such as a rectangular mesh panel). Once the plant reaches the top of the trellis, snip off the tip to encourage it to branch. Keep an eye out for red spider mites, which feed on the underside of foliage and cause leaf stippling (small white or yellow dots on leaves). Control infestations by removing affected foliage or spraying the underside of leaves with a jet of water.

# HARVESTING

Expect to start harvesting flat beans around two months after sowing. The pods are best harvested when they're still flat and the beans inside are not yet bulging. They grow quickly and prolifically, so be sure to check on your plants every few days. Save the seeds from any pods that dry on the vine to use for future crops (flat bean seeds can be saved in the same way as long bean seeds; see page 160).

# STORING

Store flat beans unwashed in a re-usable container in the fridge for up to a week. If you want to preserve the flavours of summer for longer, then you can wash, dry, chop and freeze flat beans in a re-usable freezer bag. They'll last in the freezer for up to six months.

## COOKING

When prepping flat beans prior to cooking, top and tail them like regular beans. Most flat beans are stringless; if there are any strings, then pull them off. I like to slice flat beans diagonally, as this makes the wavy edges extra pretty.

*You can use flat beans instead of long beans in the recipe for garlicky long beans with fried onion bits on page 161.*

### Flat beans in black bean garlic sauce

Here's my favourite recipe for flat beans, but long beans – or any beans – can be cooked like this as well.

Finely chop the garlic cloves. Diagonally chop the flat beans. Cut the chilli in half (if using).

Heat the vegetable oil in a wok over medium heat. Add the garlic, and stir-fry until fragrant.

Add the flat beans, and stir-fry for 2 minutes. Add the water, cover the wok with a lid, and simmer for 2–3 minutes or until most of the water has evaporated.

Remove the lid, then add the black bean garlic sauce, light soy sauce and sugar. Stir-fry for 1 minute to mix everything together.

To serve, transfer to a plate. Top with dried onion flakes and chilli (if using).

**150**

**SERVES 2**

3 garlic cloves

200 g (7 oz) flat beans

1 fresh chilli (optional)

2 teaspoons vegetable oil

¼ cup (60 ml) water

1 teaspoon black bean garlic sauce

1 teaspoon light soy sauce

½ teaspoon sugar

1 tablespoon dried onion flakes

*Black bean garlic sauce is made from fermented black soy beans (and garlic), and it adds a salty, umami flavour to dishes. You can find it at Asian grocers.*

# JAP/KENT AND KABOCHA PUMPKINS (SQUASHES)

*Cucurbita moschata* (Jap/Kent); *C. maxima* (kabocha)
Mandarin: 南瓜 – nán guā
squash • winter squash

| | | |
|---|---|---|
| PLANT TYPE | Warm-season annual | |
| PLANT FAMILY | Cucurbit family (Cucurbitaceae) | |
| PLANT SIZE | Spreading vines grow up to 2 m (7 ft) long | |
| POT FRIENDLY? | Yes, smaller varieties can be grown in a 40–50 cm (16–20 in) pot | |
| SUN | Full sun | |
| FROST | Frost sensitive | |
| WATER | Regular watering | |
| FOOD | Heavy feeder | |

I love growing pumpkins and cultivating a different kind each year – because not all of them are the same in terms of looks or flavour. These delightful vines can be left to their own devices for much of summer, and they'll reward you with a big harvest as the weather cools down. Once cured, the pumpkins can be stored well into winter and enjoyed whenever there's not much else to harvest.

Popular in Japan, **kabochas** are sweet with a dry, dense flesh that tastes like a combination of sweet potato and chestnut. They are delicious roasted, and the skins are edible too. You can also use them in curries and desserts.

Many urban gardeners shy away from growing pumpkins, as these veggies have a reputation for taking up lots of space. But I've been experimenting and found that smaller varieties (such as kabochas) can be grown vertically over an arch without the fruits breaking off.

# VARIETIES

My favourite kabocha is **Red Kuri**. I often daydream about its sweet, nutty taste. Red Kuri fruits grow to around 2 kilograms (4½ pounds), and the vine can be trained over an arch. With their deep orange skins, they hang like magical lanterns at the end of autumn.

Another great one is **Blue Kuri**. The slightly smaller fruits have a similar taste and texture to Red Kuri, but a blue-grey skin.

**Jap/Kent** is often called a kabocha but is not a true kabocha. However, it's a very popular variety with a similar flavour to the kabocha, and it has firm flesh that is drier than most other pumpkins, making it a good alternative. The fruits grow larger, up to 5 kilograms (11 pounds) in size, and the plants do best when left sprawling along the ground.

# GROWING

Pumpkins need a long, warm growing season, so it's best to start seeds indoors and transplant outside after the risk of frost has passed. Soak the seeds in warm water for a couple of hours, then sow them around 2–3 centimetres (¾–1 inch) deep. Use a heat mat if you live in a cool climate, as the seeds need warmth to germinate. Seeds should germinate within two weeks.

Pumpkins do best in full sun. Water deeply, especially when you see the fruits start to grow. Hot, dehydrated plants will drop their fruits. When watering, remember to water the soil rather than the leaves. The plants are susceptible to fungal diseases when the leaves remain wet for long periods.

Unsurprisingly, since they grow such large fruits, pumpkins are heavy feeders. So, be sure to incorporate lots of organic matter before planting, and top up when flowers start to form.

Ensure that you also have lots of pollinator-friendly flowers around your pumpkin patch. You need as many helpful insects as possible to pollinate your female pumpkin flowers, as unpollinated flowers will drop their baby fruit. It's natural for a few flowers to go unpollinated, but if all of your flowers are dropping off, then you'll have to **hand-pollinate**. This can be done by cutting off a male flower (one without a bulge on the stem underneath it) and rubbing its pollen (the yellow powder found in the centre of the flower) onto the centre (the stigma) of a female flower (one with a bulge on the stem underneath it).

**Keeping plants a manageable size** Once you have enough fruits growing on your pumpkin vine, pinch off the growing tips and any unwanted fruits. This ensures that the plant focuses its energy on growing fewer but better-quality fruits. The growing tips are edible and can be used in the same way as other greens.

In small gardens, it's best to grow one pumpkin plant but treat it well. I've found that one well-loved plant will produce more fruits than two plants feeling cramped in the same space. And remember, only the roots of the plant need to be in a garden bed. I like to plant my pumpkin in the corner of the garden bed and allow it to spread over a woodchip-mulched pathway. You can also allow your plant to scramble along the ground as a living mulch under fruit trees, or even temporarily over pavers to help suppress weeds.

# HARVESTING

Towards the end of the season, pumpkin vines will start to yellow from the oldest leaves at the base of the plant. Once this happens, it's time to see if the fruits are ready to harvest. Look for fruits that have turned their mature colour (whether that's orange or blue), then check if the closest tendril (the small squiggly bit on the vine) has turned brown and shrivelled. If both things have happened, then there's a high chance your pumpkin is ready to harvest.

Use secateurs to cut the stalk near the main vine. Leave as much of the stalk attached as possible, as this helps the fruit to store longer. If you're not confident that your pumpkin is ready to pick, then harvest one and cut it open to check. If the flesh is vibrant orange, then you can harvest the others.

Sometimes, you'll find that some fruits are ripe but others are still babies on the vine. Pick the mature ones first, then wait a few more weeks to harvest the rest.

Harvest any remaining fruits at the end of the season, even if they're small and green. Those that have started to mature on the vine will continue the ripening process on the bench. The rest of them can be used in the same way as zucchinis (courgettes).

**Curing your pumpkins** Once pumpkins have been harvested, it's best to cure them. This allows the outer skin to dry out and harden, which helps with storage.

Cure your pumpkins by leaving them on their side in a sunny spot for one or two weeks. This can be done outdoors, but I prefer to do it indoors in front of a window, so the outdoor space can be cleared for my next crop. Turn the fruits every few days so all sides get equal time in the sun.

# STORING

Once cured, separate any blemished pumpkin fruits from the rest. As much as I love imperfect produce, these ones don't store as well in the long term, so prioritise eating them first. Perfect-looking cured pumpkins can be stored in a dark, cool place on their side. I put my pumpkins on a layer of cardboard and keep them on a downstairs bookshelf. Make sure that they don't touch each other, and check and rotate them every month – compost any that turn bad. Kabochas and Jap/Kent pumpkins can be stored for up to four months.

154

*When carrying a pumpkin, don't hold it by the stalk, as it might break off. If it does, then use the fruit immediately or it will rot.*

# COOKING

Young shoots and leaves can be eaten when tender. If the shoot snaps off easily instead of bending, then it's tender enough to eat. Pumpkin seeds can be toasted for homemade pepitas!

## Chinese pumpkin cakes

These flat cakes are a popular street food in China. They're served hot, making them a perfect winter dessert. The cakes are traditionally deep-fried, but we're pan-frying here instead to save on the mess. I've used red bean paste as a filling, but you can also make the cakes without a filling. The glutinous rice flour (which is actually gluten-free!) gives the cakes a smooth but chewy texture. Despite the name 'cake', this recipe is dairy-free (I'm lactose intolerant so especially appreciate this) and vegan-friendly, too.

Cut the pumpkin into 2–3 centimetre (¾–1 inch) cubes, removing the skin in the process. Place the pumpkin into a steamer, and cook for 15 minutes or until soft.

Transfer the pumpkin to a large bowl, and mash. Add the glutinous rice flour and sugar, and mix until a dough forms. Knead the dough, then roll out into a long baguette.

Cut into 12 evenly sized pieces. Roll each piece into a ball, then gently shape into a shallow bowl. Place around 2 teaspoons red bean paste inside, then bring the edges of the dough together and pinch them closed. Place on the bench, smooth side up, and gently press flat to around 1 centimetre (½ inch) thick.

Brush the top of each cake with water, then press this side into a bowl of sesame seeds. Place the cakes onto a plate.

Heat 2 tablespoons vegetable oil in a large sauté pan over low to medium heat. Add six cakes to the pan, sesame seed side up. Cook for 4 minutes. Don't press down on the cakes – just let them be.

Flip each cake, and cook the second side for a further 4 minutes. You might see the cakes bulge in the middle while cooking – this means they're ready. Transfer the cakes to a plate.

Add another 1 tablespoon vegetable oil to the sauté pan, and cook the second batch in the same way.

Serve hot (although the cakes are just as delicious when cool). Leftovers can be stored in an airtight container in the fridge and reheated in the microwave.

**MAKES 12 CAKES**

400 g (14 oz) Jap/Kent or kabocha pumpkin (squash), skin on (or 300 g/10½ oz peeled pumpkin)

2 cups (280 g) glutinous rice flour

2 tablespoons sugar

¾ cup (190 g) red bean paste

⅓ cup (50 g) sesame seeds (white or black)

3 tablespoons vegetable oil

# LONG BEAN

*Vigna unguiculata* subsp. *sesquipedalis*
Mandarin: 长豇豆 – cháng jiāng dòu
asparagus bean • Chinese long bean • snake bean • yard bean • yard long bean

158

| | | |
|---|---|---|
| | **PLANT TYPE** | Warm-season annual |
| | **PLANT FAMILY** | Legume family (Fabaceae) |
| | **PLANT SIZE** | Climbing varieties: 2–3 m (6–10 ft) long; bush/dwarf varieties: up to 40 cm (16 in) tall |
| | **POT FRIENDLY?** | Yes, climbers in a large container with trellis; bush/dwarf varieties in a 40 cm (16 in) pot |
| | **SUN** | Full sun |
| | **FROST** | Frost sensitive |
| | **WATER** | Regular watering |
| | **FOOD** | Light feeder |

Come summer, it's bean season. Out of all the bean varieties, I love the long bean the most. It has a distinct nutty flavour that you don't find in common green beans, and the pods grow 30–70 centimetres (12–28 inches) long - you only need a few per serve! Plus, topping and tailing a few long beans is so much quicker than preparing a bunch of shorter ones.

While the long bean is part of the legume family (Fabaceae), it belongs to a different genus than the common green bean or flat bean (*Phaseolus vulgaris*) and has slightly different growing requirements. Being native to subtropical and tropical areas of Asia, the long bean needs longer summers and warmer weather than the common green bean.

If you live in a hot or tropical climate and find that common green beans become heat damaged in summer, then try growing long beans. They're much more tolerant of hot weather and dry spells. If you live in a cool climate and struggle to grow long beans, then try flat beans instead (see pages 148-50). They're more suited to cooler, shorter summers.

## VARIETIES

Long beans are mainly climbing plants that grow as vines; however, there are some bush/dwarf varieties available, too. Climbing varieties have higher yields, but bush/dwarf varieties usually mature more quickly and are a better option for cooler climates. Here are some popular varieties:

- **Black-seeded Snake** – A climbing variety, it has green pods (with black beans inside) up to 30 centimetres (12 inches) long.
- **Red Dragon** – A climbing variety, it has green pods (with red beans inside) up to 70 centimetres (28 inches) long.
- **Red Snake** – A bush/dwarf variety, it has red pods (with red beans inside) up to 30 centimetres (12 inches) long.

## GROWING

If you live in a cool climate, then consider starting seeds indoors to maximise the growing season. Sow seeds 2–3 centimetres (¾–1 inch) deep, and use a heat mat to assist with germination. Keep the soil moist, but don't overwater or the seeds might rot. It takes one to two weeks for the seeds to germinate.

Transplant your seedlings outdoors once they're around 20 centimetres (8 inches) tall and the risk of frost has passed. It's important to transplant long beans while they're small, as they grow quickly and the vines can soon become tangled. Be gentle when transplanting, as the roots are fragile and young vines can easily snap. Alternatively, direct sow long bean seeds as the weather warms up.

If you live in a warm climate, then direct sow your seeds outside – long beans actually prefer this. If your seeds don't germinate, then either the soil temperature wasn't warm enough, or the seeds rotted from too much moisture. Just try again.

Long beans thrive in warm to hot weather, so plant them in a full-sun location. They can be planted quite close together in the garden – 15 centimetres (6 inches) apart for climbing varieties, and 30 centimetres (12 inches) apart for bush/dwarf varieties.

*The seeds of both long beans and flat beans don't need to be soaked before sowing, as they don't have a hard outer shell. Soaking the seeds can actually cause them to rot.*

In cooler climates, you may find that your long beans don't do much in the first few weeks. This is normal, as they grow slowly if it's cold. They'll take off once the heat of summer hits.

If you're growing a climbing variety, then it's essential to give it vertical support. Long beans are lightweight plants, so a simple trellis is all you'll need. Once the plant reaches the top of the trellis, pinch off the growing tip to encourage it to branch out and flower. It can take two or three months for long beans to start flowering, and approximately two weeks for beans to grow after that. The beans hang in clusters and are easy to spot when harvesting.

## HARVESTING

Long beans are like most veggies – the younger they are, the better they taste. Harvest your beans before the seeds inside start to swell. Gently twist them off, and take care not to damage the buds above the beans, as plants will continue to produce more beans from the same stems. Once harvesting starts, the beans grow quickly – so you should pop outside to check on them every day.

**Saving seeds** Long beans have a very low chance of cross-pollination because the flowers self-pollinate before they bloom. So, the seeds you save will always grow into the same plants as the parent.

160

To save long bean seeds, allow a few pods to dry out on the plant. Once the pod has started to crack, break it open and collect the seeds. Spread the dried seeds on a plate indoors, and leave them for one to two weeks to completely dry. If you can't dent the seed with your nail, then it's dry. Store the seeds in an airtight jar out of direct sunlight.

Collecting the seeds means that you'll never have to buy another packet again! And you can also share them with family, friends and neighbours. Saved seeds are usually better suited to the climate in which they were grown, so they'll do better than bought seeds that may have been collected from plants grown in a different climate.

## STORING

Store long beans unwashed in a re-usable container in the fridge for up to a week. If you want to preserve them for longer, then you can wash, dry and chop the long beans and freeze them in a re-usable freezer bag – just like frozen beans in the supermarket. They'll last in the freezer for up to six months.

*Because long beans are light feeders, they're easy to grow in pots. This saves your garden space for heavy-feeding plants.*

# COOKING

Long beans can be used anywhere you'd use common green beans, only you'll notice that they have a subtle flavour difference – a nuttiness that you don't get as much from other beans. Because of the length of long beans, they can be fun to use whole. For practicality's sake, though, I usually chop them up into 5 centimetre (2 inch) pieces.

## Garlicky long beans with fried onion bits

I love to order garlicky long beans at restaurants, but for a while I couldn't figure out why my own beans never had the same nutty taste. It turned out that I was growing the wrong type of bean! (I was cultivating common green beans rather than long beans.) This recipe is vegetarian; however, it can also be made with minced pork or shrimp.

Finely chop the garlic cloves and chilli (if using). Cut the long beans into 5 centimetre (2 inch) lengths.

Heat the vegetable oil in a wok over medium heat. Add the garlic, and stir-fry until fragrant.

Add the long beans and cooking salt, and stir-fry for 10 minutes. The long beans will brown, but don't let them burn.

Add the water, and cover with a lid. Simmer for 3 minutes or until most of the water has evaporated.

Add the light soy sauce, sugar and chilli (if using). Stir-fry for 2–3 minutes or until all of the water has evaporated.

Garnish with dried onion flakes before serving.

**SERVES 2–4**

6 garlic cloves

1 fresh chilli (optional)

400 g (14 oz) long beans

2 tablespoons vegetable oil

½ teaspoon cooking salt

¼ cup (60 ml) water

1 teaspoon light soy sauce

1 teaspoon sugar

1 tablespoon dried onion flakes, to garnish

*If you find that you're only harvesting long beans very late in the season, and that you only get a short harvest window before winter comes, then it's probably because you don't have a long enough warm-growing season for long beans. If you want to grow something more productive, then try flat beans instead (see pages 148-50). These beans are equally delicious and can replace the long beans in the recipe above. They just don't need long, warm summers.*

# LOOFAH

*Luffa acutangular* (angled loofah); *L. aegyptiaca* (smooth loofah)
Mandarin: 丝瓜 – sī guā
Chinese okra • hechima • luffa

| | PLANT TYPE | Warm-season annual |
|---|---|---|
| | PLANT FAMILY | Cucurbit family (Cucurbitaceae) |
| | PLANT SIZE | 9 m (30 ft) long, spreading vines; needs vertical support |
| | POT FRIENDLY? | Yes, one plant in a 40–50 cm (16–20 in) pot, although best grown in the ground |
| | SUN | Full sun |
| | FROST | Frost sensitive |
| | WATER | Regular watering |
| | FOOD | Heavy feeder |

Have you ever used a natural shower loofah and wondered where the name came from? These sponges are made from the fibrous mature fruits of the loofah, a plant that originates from tropical Southeast Asia.

Loofah is a productive summer plant that grows a dual-purpose crop. In many parts of Asia, the fruits are harvested young and enjoyed for their flavour, which resembles that of a tender, slightly sweet zucchini (courgette).

Some fruits can be left on the vine to mature. They are then dried to become sustainable, compostable, plastic-free sponges.

I definitely recommend trying to grow your own food sponges at least once!

*When you see small loofah fruits growing, gently position them so they hang straight down – especially if you want to have straight sponges!*

# VARIETIES

There are two distinct types of loofahs:
- **Angled loofahs** have long ridges down the length of the fruits (like an okra). In Asia, these are the more common eating variety. The shape doesn't particularly lend itself to being used as a sponge, although you can try.
- **Smooth loofahs** have no ridges (like a cucumber). They're the best type for making sponges and can, of course, also be eaten.

# GROWING

Loofahs require a long growing season, so if you live in a cool climate (like I do), then you'll need to start your seeds indoors. Soak the seeds in warm water overnight before sowing, to help the seedlings crack out of their hard shells. Sow the seeds 1–2 centimetres (½–¾ inch) deep, and use a heat mat because they need a warm temperature to germinate. The warmer the temperature, the quicker they'll germinate (usually in one or two weeks).

Loofahs fruit prolifically, so you only need one plant. I like to sow a few seeds anyway, then pick the strongest plant to grow.

Once your loofah seedlings have at least three adult leaves, the weather outside has warmed up and all chance of frost is over, transplant your seedlings outside. Initially, they might grow quite slowly – don't worry, this just means that the weather is still a bit cool. Once the weather warms up, loofahs grow vigorously.

Loofah plants are heavy feeders and heavy drinkers, so be sure to enrich the soil with organic matter before planting and keep it regularly watered, especially during warm weather.

Loofahs are large vines, so set up some vertical support. One of the only difficulties with loofahs is containing their size. If you want to keep your plants small, then remove the growing tips as soon as they reach the top of your trellis. This will tell your plants to stop growing taller, and to start putting on side growth instead. Any pruned shoots and flowers can be eaten, so pop them into your next stir-fry.

Loofah flowers appear towards the end of summer. Like other cucurbits, loofah plants have both male and female flowers on the same plant. However, unlike most cucurbits, where only one flower (male or female) grows from each node, loofah plants grow multiple flowers at each node.

Male flowers grow in clusters, while female flowers grow as singles with a baby loofah attached to the stem. Usually there's only one female flower at one node, but sometimes I see two.

I typically don't hand-pollinate (see page 153) my loofahs, but it can be helpful for angled loofahs, as the flowers open in the afternoon and evening when fewer bees are around (smooth loofah flowers open in the morning, when bees are more active).

# HARVESTING

**For eating** Harvest loofahs young, one to two weeks after baby fruits start to form. You can tell if the fruit is young enough to eat by squeezing it. While loofahs are a bit softer than cucumbers, if it feels like a firm cushion, then it's young enough to eat. If it feels more like a soft sponge, then it has become fibrous and is too bitter to eat. Let that one go on and live its best life as a sponge!

**Making sponges** Leave the fruits on the vine until the plant starts to die. Then remove any loofahs that don't feel like they've developed spongy fibres. Leave the remaining mature loofahs to dry out on the vine until they turn brown and brittle. If the weather's getting wet and cold, then you can bring the loofahs inside to finish drying.

Once dried, peel off the skin and shake out the seeds. Cut into the sizes you want, then soak the loofah pieces in warm, soapy water to remove any dirt. Dry them once again in direct sunlight for a bit of natural bleaching.

**Collecting seeds** Even if you don't want to make sponges, leave one big loofah to mature on the plant so you can collect its seeds for the following year. Loofahs won't cross-pollinate with other cucurbits, and smooth loofahs won't cross-pollinate with angled loofahs (and vice versa). So, as long as you're growing only one variety of smooth or angled loofah, the seeds you save will grow into the same plant as the parent.

# STORING

Store young loofah fruits wrapped in a cloth napkin in an airtight container in the fridge for up to four days.

# COOKING

Young loofahs need to be peeled before using. Make sure that every bit of skin is removed, as it's bitter. Angled loofahs will need extra peeling compared to smooth loofahs, as you need to remove all of the angled ridges.

# Loofah, egg and capsicum stir-fry

This veggie stir-fry can be whipped up in less than 10 minutes – perfect after a long day. All it needs is three main ingredients: loofah; capsicum (sweet pepper), which usually ripens in the garden at the same time as loofah; and eggs. You can cut the loofah how you would cut a zucchini (courgette). But a common method for stir-fries is oblique/roll cutting, where you rotate the loofah as you slice it to create irregularly shaped wedges.

Peel the loofah, and slice it into small pieces. Thinly slice the garlic clove and capsicum, and chop the spring onion.

Crack the eggs into a bowl. Add ¼ teaspoon cooking salt, and beat so the yolks and whites are well combined.

In a small bowl, prepare a slurry by mixing the cornflour with 1 tablespoon water.

Heat 1 tablespoon vegetable oil in a wok over medium heat. Pour in the egg mixture, and stir-fry (using a spatula to break it into small pieces) for 1–2 minutes or until lightly cooked. Transfer to a plate, and set aside.

Heat another 1 tablespoon vegetable oil in the wok over medium heat. Add the garlic, and cook until fragrant. Add the loofah, capsicum and ¼ teaspoon cooking salt. Stir-fry for 1 minute or until tender. Add 3 tablespoons water. Stir-fry for 1–2 minutes or until only a little water remains.

Add the egg back into the wok, as well as the sugar, chicken bouillon powder and ground white pepper. Stir-fry for 1 minute to mix through. Add the cornflour slurry, and stir for about 1 minute or until thickened.

Place on a plate, and garnish with spring onion to serve.

**SERVES 2–4**

300 g (10½ oz) loofah

1 large garlic clove

50 g (2 oz) red capsicum (sweet pepper)

1 spring onion (scallion), to garnish

3 eggs

½ teaspoon cooking salt

2 teaspoons cornflour (cornstarch)

4 tablespoons water

2 tablespoons vegetable oil

¼ teaspoon sugar

¼ teaspoon chicken bouillon powder

1 pinch of ground white pepper

**165**

## MIX IT UP!

Here are two other stir-fry combinations to try:

**1. loofah + tofu + black wood ears** – All three ingredients can be chucked in the wok at the same time and stir-fried together.

**2. loofah + edamame beans + zha cai** (see page 73) – Stir-fry the edamame beans first (with a little cooking salt), then add the loofah; add the zha cai at the end of cooking.

# MALABAR SPINACH

*Basella alba*
Mandarin: 木耳菜 – mù ěr cài
Ceylon spinach • Chinese spinach • climbing spinach • Indian spinach • vine spinach

| | | |
|---|---|---|
| | **PLANT TYPE** | Perennial; grown as an annual in cool climates |
| | **PLANT FAMILY** | Basella family (Basellaceae) |
| | **PLANT SIZE** | 3 m (10 ft) long, spreading vines |
| | **POT FRIENDLY?** | Yes, grow multiple plants 10 cm (4 in) apart in a large pot with vertical support |
| | **SUN** | Full sun or part-sun |
| | **FROST** | Frost sensitive |
| | **WATER** | Regular watering |
| | **FOOD** | Medium feeder |

Malabar spinach is a fabulous heat-loving leafy green to grow in summer, when sources of leafy greens are minimal. It's a climbing perennial vine from tropical parts of Asia and, despite its name, it's not actually related to spinach. When eaten raw, it has a spinach-like flavour with a lemon tang. When cooked, it tastes very much like spinach.

Malabar spinach grows quickly and prolifically in hot weather and can provide an endless supply of greens. It's a low-fuss plant, too. I often let mine just ramble through the garden without giving it much attention. It grows as a perennial in warm climates; cool-climate gardeners only really need to sow it once, as it easily self-seeds in the same spot year after year.

*Malabar spinach is easy to propagate and share with friends. All you need to do is cut off a 20-30 centimetre (8-12 inch) section of the growing tip, remove the bottom leaves, pop it in water and let it root.*

## VARIETIES

**Basella alba** is the most common and has green leaves and stems. But there's also a red cultivar, **B. alba 'Rubra'**, which has vibrant red stems and green leaves, and is beautiful in the patch. This plant produces purple berries after flowering, which can be used as a natural dye. Take care, though, as they will stain everything!

## GROWING

Malabar spinach is easy to direct sow, but the seeds need a warm temperature to germinate. First soak the seeds overnight to help speed up germination, then pop them into any gaps you have in the garden, as long as there is around 10 centimetres (4 inches) of space around them. Seedlings will grow slowly in cool weather. However, the moment that the weather warms up, they'll be straight off to the skies.

These leafy vines can grow up to 3 metres (10 feet) long, so it's best to provide vertical support, even if it's just a couple of long stakes. This also helps to keep the leaves clean and off the ground. Malabar spinach is a prolific grower and will consistently send out leaves throughout summer, so two or three plants are more than enough for a backyard patch.

The plants are thirsty and prefer to be regularly watered. The more you water, the faster they'll grow. They are drought tolerant, but a reduction in water can trigger flowering, shortening the harvest window.

## HARVESTING

Harvest your Malabar spinach by snipping off young, tender parts of the vine throughout the growing season. Doing this will also encourage it to grow bushier. When the vine reaches the top of your trellis, snip off the growing tip to make it grow new shoots below.

Once the plant starts to flower, the leaves will turn bitter. The small, succulent-like flowers turn into large, round seeds. These seeds drop off, sit dormant in the soil and naturally resprout when the weather warms up the following year. Simply transplant the new seedlings to wherever you want them. If you don't want the plant to self-sow, then cut it down as soon as it flowers.

## STORING

Store fresh Malabar spinach leaves wrapped in a cloth napkin in an airtight container in the fridge for up to five days. If you have too much Malabar spinach, then you can blanch and freeze it for later use. These frozen leaves can be popped straight into dishes as you're cooking them.

# COOKING

The thick and succulent mature leaves of Malabar spinach have a mucilaginous (slimy) texture when cooked. This makes them best suited to soups, stews and curries, where the texture is easily masked and can help to thicken dishes. Young shoots and leaves are less mucilaginous, so use these in stir-fries. Very young leaves can be added to salads and eaten raw.

## Build a curry

When I have a basketful of veggies, I often make a curry with them. It's such an adaptable dish, and you really can use anything. I always start a curry with the core ingredients of garlic, onion, curry paste/powder and protein, then add in whatever is in season from the garden. I usually choose a ratio of one leafy veggie, one flavour soaker and one other seasonal veggie. Malabar spinach is one of my favourite leafy greens to add to a summer curry, as it's fleshy and soaks up the flavours so well. Use the table below to inspire your next curry.

168

| CORE INGREDIENTS | SEASON | LEAFY VEGGIES | FLAVOUR SOAKERS | SEASONAL VEGGIES |
|---|---|---|---|---|
| • Garlic<br>• Onion<br>• Curry paste/ powder of your choice<br>• Protein – beans, beef, chicken, chickpeas, fish, lentils, pork, prawns, tofu | SUMMER | • Malabar spinach<br>• Sweet potato leaves | • Eggplant (aubergine)<br>• Hairy melon<br>• Loofah<br>• Winter melon | • Baby corn<br>• Beans<br>• Capsicum (sweet pepper)<br>• Mushroom |
| | WINTER | • Bok choy<br>• Celtuce leaves<br>• Chinese broccoli<br>• Wombok | • Cauliflower<br>• Daikon<br>• Dried beans<br>• Cured shark fin melon | • Carrot<br>• Mushroom<br>• Snow pea (mange tout)<br>• Sweet potato tuber |

# SHARK FIN MELON

*Cucurbita ficifolia*
Mandarin: 鱼翅瓜 – yú chì guā
angel's hair melon • chilacayote • fig-leaf gourd • fig-leaf squash •
Malabar gourd • Siam squash

| | PLANT TYPE | Perennial; grown as an annual in cool climates |
|---|---|---|
| | PLANT FAMILY | Cucurbit family (Cucurbitaceae) |
| | PLANT SIZE | 5–15 m (16–49 ft) long, spreading vines |
| | POT FRIENDLY? | Perhaps in a 50 cm (20 in) pot with a large trellis, but it's best grown in the ground |
| | SUN | Full sun |
| | FROST | Frost sensitive |
| | WATER | Regular watering |
| | FOOD | Heavy feeder |

Shark fin melon gets its name from its stringy, noodle-like flesh, which is said to look like shark fin when cooked. Its leaves are shaped like those of figs, which has inspired two other names: fig-leaf gourd and fig-leaf squash.

It's a prolific plant that can grow 50+ fruits on a happy, sprawling vine. The melons are green with white speckles, resemble watermelons and weigh up to 6 kilograms (13 pounds) each. They even have small black seeds inside, although these look more like pepitas (pumpkin seeds) and can be eaten like them, too.

Shark fin melon originates from Central and South America, where it's often used to make desserts. In Asia, it's more commonly used in savoury dishes, especially soups. The melon has a pleasantly fresh, light taste - similar to a winter melon - and can pick up the flavours of whatever it's cooked with.

*Shark fin melon won't cross-pollinate with other cucurbits in your garden. This means that you can save the seeds from your best-producing fruit and know that they'll grow true to the parent plant the following year.*

# GROWING

Sow shark fin melon seeds 1.5 centimetres (⅓ inch) deep. Seeds take one to two weeks to germinate; they need warmth, so use a heat mat. Transplant seedlings outdoors after the risk of frost has passed.

Shark fin melon is a large, vigorous vine, so give it as much space as you can; if the plant gets too enthusiastic, then you can prune it back to keep it contained. A good idea is to grow it over an arch, pergola or trellis.

You can also let the plant sprawl over the ground. It will happily spread, and can double as an edible ground cover.

Keep your shark fin melon plant well-watered until it's established. Once established, the plant will be quite drought tolerant; however, it's best to regularly water it if you want it to grow more fruits. The plant has both male and female flowers, which are pollinated by bees, and will start to fruit from late summer onwards.

Prune throughout the growing season. Young shoots, leaves and flowers that are pruned off can all be eaten, so remember to keep those bits for dinner that night.

Shark fin melon is more resistant to powdery mildew than other cucurbits, which makes it a great option if you live somewhere with a humid climate.

# HARVESTING

Shark fin melons are usually harvested mature and eaten throughout winter. However, they can be harvested at any time during the growing season. Young melons have thin skins and small seeds, so the whole thing can be eaten just like a zucchini (courgette).

For mature melons, wait until the end of the season when the vine leaves start to yellow. Harvest the melons by cutting them off the vine with the entire stem attached. Next, cure the fruits by placing them in a ventilated space with indirect light. This helps to harden and dry the skin for long-term storage.

# STORING

Store cured shark fin melons in a dark, ventilated space at 10–16 degrees Celsius (50–61 degrees Fahrenheit), and make sure that no fruits are touching. Prioritise eating the imperfect fruits first; rotate the others regularly to reduce the chance of rot, and compost any that don't make it. Cured shark fin melon is known for its incredibly long storage life – longer than winter melon and pumpkin (squash) – and can be kept for up two years. It's the perfect crop for bulking up your winter pantry.

# COOKING

If you grow and store cured shark fin melons, then you can always count on having some on hand when garden harvests are lean. Cured fruits can replace loofah, winter melon and zucchini (courgette) in recipes. They can be roasted, and used in soups, stews and stir-fries. When cutting open a cured shark fin melon, be extra careful. It has a tough skin (which helps with storage), so use a sharp knife and go slowly.

## Shark fin melon soup

Here's a light, refreshing clear soup to make with your shark fin melon (you can also use winter melon; see pages 186–8). Like most cooking in my kitchen, it's very much a choose your own adventure. If you don't have jujubes or goji berries, then you can leave them out or replace them with a few raisins. The seafood is there for umami goodness. If you're vegetarian or vegan, then use black wood ears and/or kombu seaweed and veggie stock instead. Serve the soup with a bowl of rice for an easy weeknight dinner.

Prepare the shark fin melon by removing the skin and seeds, then cutting it into 3–4 centimetre (1–1½ inch) cubes. Slice the carrot and sweet corn into chunks. Quarter the shiitake mushrooms. Rinse the dried scallops/shrimps and jujubes.

Place the shark fin melon, carrot, sweet corn, shiitake mushrooms, scallops/shrimps, jujubes and water into a large saucepan. Bring to a boil, then simmer over low heat for 30 minutes. Meanwhile, chop the spring onion.

Add the goji berries, vegetable oil, chicken bouillon powder, cooking salt and ground white pepper. Stir, and cook for a further 1 minute. Serve in bowls, and garnish with spring onion.

172

**SERVES 4**

600 g (21 oz) shark fin melon (or half a medium-size one)

1 carrot

1 sweet corn

4 shiitake mushrooms

30 g (1 oz) dried scallops and/or dried shrimps

6 dried jujubes

4 cups (1 litre) water

1 spring onion (scallion), to garnish

2 teaspoons goji berries

1 teaspoon vegetable oil

½ teaspoon chicken bouillon powder

1 teaspoon cooking salt

1 pinch of ground white pepper

## OTHER WAYS TO USE SHARK FIN MELON

Use the stringy flesh in the same way as zucchini (courgette) noodles, or add it to noodle dishes for an extra layer of texture. Harvest the top 10 centimetres (4 inches) of the plant's growing tips, as they are the most tender, and add them to stir-fries, soups or salads.

Mature shark fin melons can contain more than 500 seeds, which can be tossed with olive oil and cooking salt, roasted and eaten like pepitas (pumpkin seeds). Note that sprouting seeds are toxic – don't eat those!

# SWEET POTATO AND ITS LEAVES

*Ipomoea batatas*
Mandarin: 番薯 – fān shǔ
keledek • kumara

| PLANT TYPE | Perennial; grown as an annual in cool climates |
| --- | --- |
| PLANT FAMILY | Morning glory family (Convolvulaceae) |
| PLANT SIZE | 3 m (10 ft) long, spreading vines |
| POT FRIENDLY? | Yes, one plant in a 30–50 cm (12–20 in) pot |
| SUN | Full sun |
| FROST | Frost sensitive |
| WATER | Regular watering |
| FOOD | Light to medium feeder |

The unassuming sweet potato is actually a dual-purpose crop. In Western cultures, only the tubers (the potatoes) are eaten. But across Asia, both the sweet potato and its leaves are enjoyed. Sweet potato leaves offer a generous source of tasty greens in summer, when most other leafy greens aren't in season. It's one of the reasons I love growing sweet potato plants.

The leaves are especially prized because they're extremely nutritious (they're high in vitamin C and B6) and incredibly delicious - they taste like regular spinach crossed with water spinach. They are a great summer substitute for spinach but contain less oxalic acid (see page 130). However, I rarely see sweet potato leaves in Asian grocers where I live. So, if you want to enjoy them, you'll likely have to grow your own.

## VARIETIES

Sweet potato varieties can be categorised based on flesh colour. Leaf shapes also differ across varieties – some are heart-shaped, while others look like maple leaves.

**Orange-fleshed sweet potatoes** are the most common type in Australia. These grow more quickly than other varieties, making them ideal for temperate and cool climates, especially if you want to harvest the tubers.

**White- and purple-fleshed sweet potatoes** also exist – you'll often see them in Asian grocers. These grow more slowly than orange-fleshed varieties and are more suited to a warm climate. You can still grow them in temperate and cool climates for their leaves, although a reasonable tuber harvest can't be guaranteed.

## GROWING

To start a sweet potato plant, you'll need a slip: a cutting that has been grown from a sweet potato tuber. You can grow your own, or buy them already potted up from a nursery. If you live in a cool climate, then it's best to start your slips indoors in winter to get a head start.

**Cultivating your own slips** Sweet potatoes have two ends: a pointy end and a blunt end. The pointy end is where roots form and needs to be under water. The blunt end is where slips form and needs to face upwards. (There are exceptions, though – some varieties grow both roots and slips on the same end.) Suspend your sweet potato in a jar of water with the pointy end down, using a few toothpicks to hold it in place.

Keep your jar in a spot with bright, indirect light, and change the water every few days to keep it fresh. Slips grow faster in warm weather, so if you live in a cool climate, place your jar on a heat mat to speed up the process.

In a few weeks, you'll see roots growing from the bottom, followed by little sprouts (slips) on top. Once the slips have three or four fully grown leaves, gently snap off the slips. Remove the bottom leaves (so two or three are left on top), and place the slips in a jar of water with just the stem submerged (not the remaining leaves). The slips will start to root after a week or so. When a few roots have formed, the slips are ready to pot up or transplant into the garden. If you're in a cool climate, then wait until the last frost has passed before planting them outside. If you're in a warm climate, then you can plant straight into the soil.

**In the garden** Grow your sweet potato plant in a warm and sunny spot, and add organic matter to the soil before planting. Sweet potatoes require a long growing season to form a good number of tubers. In warmer climates, you'll be able to grow both sweet potato leaves and tubers. In cooler climates, you'll get a fabulous harvest of leaves but fewer or smaller tubers, depending on the variety you grow.

As sweet potato plants grow, they start to shoot out multiple long vines. If you want to harvest both leaves and tubers, allow the vines to sprawl across the ground. This encourages roots to grow from the nodes along the vines. These extra roots help the plant to gather more nutrients from the soil, resulting in better growth and productivity. As a further surprise, they might produce tubers where they root as well, although they will be smaller than the main batch that grows at the base of the plant.

If you want to grow sweet potatoes only for their leaves, then train them up a trellis. This saves space and keeps the leaves clean and off the soil. However, these vines won't be able to send extra roots into the soil, which reduces the quality of tubers you'll get.

## HARVESTING

176

You can harvest sweet potato leaves at any time throughout the season. Harvest by either cutting off all of the young shoots, or cutting off the shoots and leaves from only one section of the vine. Sweet potato vines produce a sticky white sap when cut, so scissors are a great idea.

The actual tuber harvest is done at the end of the growing season, as the weather cools down, right before your first frost. Wait for a period of dry weather, and stop watering your plant for a couple of weeks before harvesting. This way, the ground will be less muddy and the tubers easier to dig up. Harvest the sweet potatoes by very gently digging around the base of the plant with your hands. Use a garden fork to loosen the soil around the plant base if necessary. Take care, as the tubers are fragile and easily damaged.

**Curing sweet potatoes** Once your sweet potatoes are harvested, leave them outdoors to dry out in the sun for half a day. Then place them in an open cardboard box in a hot (30 degrees Celsius/ 86 degrees Fahrenheit), humid but well ventilated spot indoors for one to two weeks to let them cure. During this time, the starches will convert to sugars, making them sweeter. The skin will also dry out and harden, helping them to store longer.

## STORING

Prioritise eating the imperfect sweet potato tubers first, as the perfect ones will store for much longer. Keep cured sweet potatoes in a cool, dark place for up to three months. Store sweet potato leaves wrapped in a cloth napkin in an airtight container in your fridge for up to five days.

## COOKING

Young sweet potato leaves can be eaten raw in salads, while older ones can be cooked in stir-fries. With the tubers, I love to simply steam them – it really brings out the fluffy texture and rich, sweet flavour of homegrown sweet potatoes. Gently wash the tubers, and slice them into 2–3 centimetre (¾–1 inch) pieces (skins on). Bring a saucepan of water to a boil, then steam the pieces for 10 minutes or until soft and tender. Once steamed, the skins will easily slip off the tubers as you eat them.

*Sweet potato leaves are not to be confused with regular potato leaves. Sweet potato leaves are delicious, while regular potato leaves are poisonous!*

# Shanghai fried noodles

I like to add sweet potato leaves to Shanghai fried noodles, which are typically eaten on birthdays because long noodles symbolise a long life. Luckily, my birthday is in summer when sweet potato leaves are at their peak!

Thinly slice the mushrooms and brown onion. Rinse and drain the mung bean sprouts.

Bring a large saucepan of water to a boil. Add the noodles and cook for 30 seconds, then drain and rinse in cold water to help prevent the noodles from clumping together. Set the noodles aside.

Crack the eggs into a bowl. Add the cooking salt, and beat so the yolks and whites are well combined.

Heat ½ tablespoon vegetable oil in a wok over medium heat. Pour in the egg mixture, and stir-fry (using a spatula to break it into small pieces) for 1 minute or until lightly cooked. Transfer to a bowl, and set aside.

Heat 1 tablespoon vegetable oil in the wok over medium heat. Add the mushrooms, and stir-fry for 2 minutes or until fragrant and cooked through. Add the onion, and stir-fry for 1 minute or until fragrant. Add the mung bean sprouts, and stir-fry for 30 seconds or until lightly cooked through.

Add the noodles to the wok, then add the light soy sauce, dark soy sauce, white vinegar, sugar and ground white pepper. Mix together.

Add the sweet potato leaves, prawn meat (if using) and egg mixture. Stir-fry for 3 minutes or until the sweet potato leaves have wilted. Stir in the sesame oil before serving.

**SERVES 3–4**

4 shiitake or button mushrooms

1 brown onion

150 g (5¼ oz) mung bean sprouts (to grow your own, see page 199)

500 g (18 oz) thick Shanghai noodles (or Hokkien noodles)

3 eggs

¼ teaspoon cooking salt

1½ tablespoons vegetable oil

1 tablespoon light soy sauce

3 teaspoons dark soy sauce

1 tablespoon white vinegar

1 teaspoon sugar

1 pinch of ground white pepper

200 g (7 oz) sweet potato leaves

100 g (3½ oz) cooked and shelled prawn meat (optional)

1 teaspoon sesame oil

*Sweet potato leaves are also delicious in a stir-fry on their own. Follow the water spinach recipes on page 183, but replace the water spinach with the same weight of sweet potato leaves.*

# WATER SPINACH

*Ipomoea aquatica*
Mandarin: 空心菜 – kōng xīn cài
kang kong • morning glory • ong choy • swamp cabbage • water convolvulus

| | | |
|---|---|---|
| 🏷️ | **PLANT TYPE** | Perennial; grown as an annual in cool climates |
| 🌱 | **PLANT FAMILY** | Morning glory family (Convolvulaceae) |
| ⤢ | **PLANT SIZE** | 2–3 m (7–10 ft) long, spreading vines; can also be grown upright in a pot |
| 🪴 | **POT FRIENDLY?** | Yes, grow multiple plants in one medium pot |
| ☀️ | **SUN** | Full sun or part-sun |
| ❄️ | **FROST** | Frost sensitive |
| 💧 | **WATER** | Consistently wet – lives in water |
| ✺ | **FOOD** | Heavy feeder |

Among all Asian greens, one stands out for its satisfying crunch: water spinach, a leafy green that thrives in warm weather. It has a mild and distinct taste that's a little sweet and nutty. Both the stems and leaves are edible, but I find the crispy, hollow stems especially delicious. Water spinach is best eaten fresh, so I make sure to have some in my garden during summer so we can enjoy it while it's in season.

Water spinach is a tropical plant, native to ponds and swamplands in Southeast Asia. The plant looks like miniature bamboo. In warm climates, it thrives as a perennial, growing vigorously and even invasively. In cool climates, it is more commonly grown as an annual because of its frost-tender nature. Although it's semi-aquatic, you don't need a pond to grow it – I'll show you how you can easily cultivate it in any backyard.

# GROWING

The easiest way to grow water spinach is to start from cuttings. Plants grown from cuttings grow more quickly than those grown from seed. Here is a step-by-step guide to growing water spinach from cuttings:

1. Buy a bunch of water spinach from an Asian grocer.
2. Remove the bottom few leaves (use them for dinner that same night), and pop the bunch in a jar of water, ensuring that at least two nodes are under water.
3. Leave the jar on a windowsill with indirect light, and change the water every two days.
4. Roots will start to develop within a week.

I'd only suggest growing from seed if you're not able to find water spinach at an Asian grocer. This is how to grow water spinach from seed:

1. Soak the seeds overnight.
2. Plant the seeds 1 centimetre (½ inch) deep in punnets of seed-raising mix.
3. Keep the seeds consistently moist until they germinate.
4. Use a heat mat, as water spinach seeds need a warm temperature to germinate.

Once your water spinach cuttings grow roots, or your seedlings have three or four adult leaves, you can transplant them into their final growing spot. If you have a pond or a spot in your garden where water always pools, then that's the best place for water spinach. Otherwise, you can create a small pond in a tub (see page 182).

Water spinach needs warm weather and a location in full sun or part-sun to grow well. In cooler climates, pop it next to a brick wall that receives afternoon sunlight to help extend the season. It also needs to be kept continuously wet, which is why a pond in a tub is a great option. You can't really overwater a water spinach plant, but you can under water it. Water spinach is a hungry plant, so add a nitrogen-rich liquid fertiliser to the water reservoir every couple of weeks to keep it happy. The plants are largely free of pests.

*In warmer climates, don't let water spinach grow in or near open bodies of water (such as lakes) because it can spread and become invasive. Some parts of the United States have strict regulations about growing this plant because of its invasive nature, so double-check before getting started.*

**WATER SPINACH**

## HARVESTING

Harvest water spinach by cutting close to the base, leaving at least two nodes behind. New shoots grow quickly from these nodes, and you'll be able to harvest again in just a few weeks' time.

On hot days, it's best to harvest in the morning when plants have the most moisture in them. This prevents them from wilting too fast.

## STORING

Once harvested, keep the water spinach leaves fresh by storing them in the fridge, with their stems in a jar of water; however, they won't keep for too long so are best eaten quickly.

182

### DO-IT-YOURSELF POND IN A TUB

If you don't have a natural pool in your garden, then why not create your own small pond? You'll need:

- numerous pots, 20 centimetres (8 inches) high and wide, with drainage holes
- quality potting mix
- numerous water spinach plants
- 1 large UV-resistant tub, at least 10 centimetres (4 inches) high.

Fill the pots with potting mix. Plant two or three water spinach plants in each pot. Place all of the pots into the tub, and fill the tub with water. Every week or so, drain the tub and add fresh water (especially if the water starts to look or smell foul).

10 cm
(4 in)

UV-resistant tub

# COOKING

Because water spinach has a unique texture, I enjoy it simply on its own and rarely stir-fry it with other ingredients. The stems are really the hero of the veggie, so use them all. Here are two different ways I love to cook water spinach.

## Sambal kang kong

This is a very popular Malaysian dish, and I love how the textures and flavours work together. At home, I make a basic version using homemade sambal.

Chop the garlic. Chop the water spinach into 5 centimetre (2 inch) pieces.

Heat the vegetable oil in a wok over medium heat. Add the garlic, and stir-fry until fragrant.

Add the water spinach, sambal and light soy sauce, and stir-fry for 2–3 minutes or until the water spinach has wilted. Serve warm.

**SERVES 2**

2 garlic cloves

200 g (7 oz) water spinach

1 tablespoon vegetable oil

1–4 teaspoons sambal, to taste (to make your own sambal, see page 233)

1 teaspoon light soy sauce

## Water spinach in garlic

For a spice-free alternative, try this garlic version – it's my other go-to recipe for water spinach and really allows the veggie's natural flavours to shine.

Finely chop the garlic. Chop the water spinach into 5 centimetre (2 inch) pieces.

Heat the vegetable oil in a wok over medium heat. Add the garlic, and stir-fry until fragrant.

Add the water spinach and cooking salt, and stir-fry for 2–3 minutes or until the water spinach has wilted. Serve warm.

**SERVES 2**

3 garlic cloves

200 g (7 oz) water spinach

1 tablespoon vegetable oil

¼ teaspoon cooking salt

# Water babies

If you love growing edible water plants, then here are some more Asian favourites to try in your garden.

## WASABI

*Eutrema japonicum* (syn. *Wasabia japonica*)
Mandarin: 山葵 – shān kuí
Japanese horseradish

Wasabi is a very challenging but rewarding plant to grow. It has very specific growing condition requirements – a full-shade spot, flowing water, lots of humidity and a cool temperature – and it will be two years before you can harvest it. This compact plant (small enough that it can be grown in a pot) is grown for its rhizome, which is grated to make a strong, spicy paste that's served with sushi and sashimi.

# WATER CHESTNUT
*Eleocharis dulcis*
Mandarin: 菱角 – líng jiǎo
Chinese water chestnut

Water chestnuts are easy to grow, but they do need a bit of space as they spread – try a large pot, an old bathtub or a clamshell pool. For good-sized chestnuts, the plants will need full sun. The plants grow through the warmer seasons, and the chestnuts are harvested in autumn when the weather cools down. With their sweet, nutty flavour and crunchy texture, the chestnuts are great in dumpling fillings, stir-fries and soups.

*Plant water chestnuts in moist soil, and when they are 30 centimetres (12 inches) tall, flood the container with water so that the water level is 20 centimetres (8 inches) above the soil.*

# WATERCRESS
*Nasturtium officinale*
Mandarin: 西洋菜 – xī yáng cài
yellowcress

This easy-to-grow, frost-hardy perennial requires a partly sunny position and can be grown by a window indoors. Highly nutritious and commonly used fresh as a 'cut and come again' green, watercress has a spicy flavour and works well in salads or as a garnish. It can also be grown as microgreens.

# WINTER MELON

*Benincasa hispida*
Mandarin: 冬瓜 – dōng guā
ash gourd • ash melon • Chinese preserving melon • tallow gourd •
wax gourd • white gourd • winter gourd

| | | |
|---|---|---|
| 🏷 | PLANT TYPE | Warm-season annual |
| 🌿 | PLANT FAMILY | Cucurbit family (Cucurbitaceae) |
| ⤢ | PLANT SIZE | Spreading vines up to 6 m (20 ft) long; needs vertical support |
| 🪴 | POT FRIENDLY? | Yes, one small-variety plant in a 40–50 cm (16–20 in) pot |
| ☀ | SUN | Full sun |
| ❄ | FROST | Frost sensitive |
| 💧 | WATER | Regular watering |
| ⚬ | FOOD | Heavy feeder |

Winter melon is a large gourd. Whenever I think of the word 'gourd', what comes to mind is fleshy veggies with no taste. Not winter melon - it's the golden star of the lot. It has the crisp texture and mild, refreshing flavour of watermelon rind. To me, it tastes like a fruit even though it's used as a veggie, which makes it versatile enough for both sweet and savoury recipes. When cooked, it picks up the flavours of other ingredients in the dish.

Despite its name, winter melon grows in summer and is harvested when the weather cools down. Mature winter melons develop a waxy outer coating that protects the fruit, giving them a long storage life so they can be enjoyed through winter - hence the name. This helps you to spread out your summer harvest, so you can enjoy it during the cooler seasons when there's less going on in the garden.

*Winter melon flowers are edible and can be treated like zucchini (courgette) flowers. Young shoots can be cooked in stir-fries or eaten raw in salads.*

# VARIETIES

Winter melon varieties differ in shape and size. Some are round, while others are long. The biggest ones can grow fruits that weigh up to 20 kilograms (44 pounds); the smaller ones have fruits that weigh a tenth of that.

It's always a thrill to grow larger varieties, but it's more practical to grow smaller ones. Varieties with small to medium fruits are best, as they're easier to manage in the kitchen.

# GROWING

If you live in a cool climate, then it's best to start your seeds indoors so they have a long growing season of four to five months. The seeds have a hard shell, so soak them in warm water overnight before sowing. They require a very warm temperature to germinate, so a heat mat is essential.

Transplant the seedlings into a full-sun position in the garden once they have three or four adult leaves and the weather has warmed up. For small to medium varieties, use a large, strong trellis to support the long vines. For large varieties, let the vines sprawl across the ground. I like to grow my winter melons over a reo mesh arch, which supports fruits up to 7 kilograms (15½ pounds) – and probably heavier – with no problems.

Water your winter melon plants regularly, and keep the soil moist during dry spells in summer. Be sure to water the soil and not the leaves, as the plants are susceptible to powdery mildew.

Plants grow both male and female flowers. Once flowers start blooming, top up the soil with extra organic matter. Consider pruning any excess vine growth, so only three or four fruits are allowed to mature. This helps the plant to focus its energy on growing a few large melons, rather than many small ones. Once baby fruits form, it takes around two months for them to mature.

# HARVESTING

Winter melon is typically grown for its mature fruits, although you can harvest them at any stage of growth. I harvest mine when they're mature, as I have plenty of other veggies to enjoy during summer.

To harvest mature winter melons, wait until the vines have started to yellow and die down. Cut the stem, but leave as much of it attached to the melon as possible to help keep it fresh. Be prepared to catch the melon once you cut the stem, as it can be heavier than it looks! Mature winter melons will either have a waxy coating or develop it soon after harvest.

# STORING

Mature melons can be stored for five months or more in a dark, dry place with a temperature of 13–15 degrees Celsius (55–59 degrees Fahrenheit). Store unblemished melons only, and don't wash the waxy coating off the fruits until you're ready to eat them.

## COOKING

Prepare winter melons by washing them to remove the waxy outer coating. Once cut open, the fruits keep for only a week. This is why it's more practical to grow smaller winter melons – you can easily consume them before they go off. However, if you've cut open a large winter melon with no hope of finishing it quickly, then here are a couple of ways to sort it out:

1. Cut up the flesh into medium-sized cubes, divide them into zip-lock bags and freeze. These cubes can be plopped straight into soups and curries.
2. Chop up and share the melon with neighbours and friends.

Winter melon is usually used in savoury cooking, such as soups or curries. (To make winter melon soup, follow the shark fin melon soup recipe on page 172, but swap one melon for the other.) However, the tea recipe below is a sweet and unique way to use lots of winter melons in one go.

## Winter melon tea

This tea is traditionally enjoyed cold in summer but can also be served hot. The recipe produces a concentrated syrup that you can keep in the fridge for up to a week. Simply dilute one part syrup to five parts water (or to taste). You can turn the diluted drink into boba/bubble tea by adding your choice of ingredients: lychee jelly, pearls, milk, black tea, ginger, and so on. Alternatively, freeze the syrup in ice-cube trays, then simply pop a block or two into water for iced winter melon tea. To store the syrup, you'll need a small clean and sterilised glass jar with a lid (see page 72).

Wash the winter melon to remove the waxy coating, then slice it into 2–3 centimetre (¾–1 inch) cubes, leaving the skin and seeds intact.

Place the winter melon cubes into a large saucepan, and stir in the brown sugar until it has dissolved. Set aside for 2 hours so that water is extracted from the winter melon.

Bring the saucepan of winter melon to a boil. Reduce the heat to low, and simmer for 1–2 hours or until the winter melon has turned mushy and transparent. Stir now and then to prevent the winter melon from burning. You can keep the saucepan covered or uncovered. Keeping the lid on creates a watery syrup; keeping the lid off (my preference) creates a concentrated syrup that needs to be diluted more.

Allow the syrup to cool, then pour it through a cheesecloth or strainer and into the jar. Place the lid on the jar, and store the jar in the fridge for up to 1 week.

188

**MAKES APPROX. 200 ML (6¾ FL OZ) SYRUP**

1 kg (2¼ lb) mature winter melon (scale up according to how much winter melon you have)

250 g (9 oz) brown sugar (it gives the tea its colour and caramel flavour)

# More magical melons

If you like growing winter melon, then here are two other Asian melons to try. Both are grown in largely the same way as winter melon. I've noted any differences below.

## BITTER MELON
*Momordica charantia*
Mandarin: 苦瓜– kǔ guā
balsam pear • bitter apple • bitter gourd • bitter squash

Bitter melon is one of those polarising veggies that you either love or hate. It tastes like a bitter cucumber and is best harvested very young, as the older the fruit is, the more bitter it becomes. However, like eggplant (aubergine), you can soak sliced bitter melon in water before cooking with it, to help reduce its bitterness a little. It's typically used in stir-fries, curries and soups.
Bitter melon is a perennial (often grown as an annual), and it has ornamental leaves and exquisite-looking fruits. The plant is largely free of pests – most wildlife prefer to eat something else!

## HAIRY MELON
*Benincasa hispida* var. *chieh-gua*
Mandarin: 毛瓜 – máo guā
fuzzy gourd • fuzzy melon • fuzzy squash • hairy gourd • small winter melon

If you live in a cool climate or you don't have a long enough growing season for winter melon, then try hairy melon. It's a specific variety of winter melon that's grown for its abundant immature fruits and because it grows more quickly than other winter melon varieties. You'll need to rub the hairs off the fruits before cooking them.
If you let the fruits mature on the vine, then they'll lose their fuzzy exterior and develop a waxy coating just like other winter melon varieties. Once mature, they're referred to as winter melons, and they can be stored and enjoyed just like regular winter melons. (When winter melon is harvested young, it's often called hairy melon!)

# YEAR-ROUND

# VEGGIES

# CHOY SUM

*Brassica rapa* var. *parachinensis*
Mandarin: 菜心 – cài xīn
Chinese flowering cabbage • you cai • yu choy • yu choy sum

| | | |
|---|---|---|
| 🪴 | **PLANT TYPE** | Warm- or cool-season annual, depending on variety |
| 🌿 | **PLANT FAMILY** | Mustard family (Brassicaceae) |
| ⤢ | **PLANT SIZE** | 20–40 cm (8–16 in) tall, around 20 cm (8 in) wide |
| 🪴 | **POT FRIENDLY?** | Yes, plant three in a 30 cm (12 in) pot |
| ☀ | **SUN** | Full sun or part-sun |
| ❄ | **FROST** | Depends on variety |
| 💧 | **WATER** | Regular watering |
| ❖ | **FOOD** | Medium feeder |

Choy sum is a popular, well-loved staple across many parts of Asia. It's also *the* fastest-growing leafy green in this book – yep, even faster than bok choy! There are varieties for both warm and cool seasons, so pop in a plant anytime you have an unexpected gap in your veggie patch to obtain a bonus harvest. Choy sum is great for making the most of the permaculture principle of using the edges and valuing the marginal spaces.

Choy sum is commonly grown for its tender flowering stem, but the greens are edible as well. It's sweet, tender and crisp because it's harvested so young. My mum loves its thin stems because they cook more quickly than bok choy. The best bits, however, are the small yellow flower buds, which are tender and sweet. Choy sum is delicious on its own - serve it steamed or boiled, or drizzled with sauce. It's also great in stir-fries and soups.

## VARIETIES

Most green-stemmed and green-leafed varieties prefer warm temperatures and are best grown as warm-season annuals, as they are not frost tolerant at all. **Early Green** needs only 30–40 days to grow, while **Jung Green** and **Late Green** are more heat tolerant so are better for a warm climate. These are best sown in early spring, late summer or early autumn, depending on your climate.

**Hon Tsai Tai**, a purple-stemmed variety, is the only one that tolerates frost and prefers slightly cooler temperatures than others. This variety is best planted as a cool-season annual in autumn.

## GROWING

In general, choy sum plants are sensitive to temperature. They grow best when it's 15–25 degrees Celsius (59–77 degrees Fahrenheit). Temperatures above or below this range increase the risk of bolting.

Choy sum's tendency to bolt is not exactly a bad thing, given that the flower buds are the most prized part of the plant. However, it's important to keep in mind the temperature requirements, as your choy sum plants need enough time to grow thick stems and lush leaves before they flower. Otherwise, you'll be harvesting very small plants (and flower buds). The good news is that choy sum has a short life cycle – as little as 30–40 days – from seed to harvest. So, you only need to ensure that the preferred temperature is maintained for a short period of time.

Choy sum grows fast, so if you have the space and your soil temperature suits, then you might as well direct sow the seeds. Otherwise, raise choy sum in punnets as you would other Asian greens, then transplant seedlings when they have three or four adult leaves.

Choy sum plants are small, so space them 20–30 centimetres (8–12 inches) apart. They prefer fertile, well-drained soil, as they don't cope with waterlogged roots.

*The Mandarin name for choy sum, cài xīn, actually translates to 'vegetable heart', referring to the inner flowering part of the veggie.*

*Confusingly, though, the flowering centres of bok choy (and other Asian greens) are also referred to as cài xīn because it's used as a descriptive term as well as the Mandarin name for choy sum.*

## HARVESTING

Depending on the variety, your choy sum will start flowering four or five weeks after sowing. Once you see the first few yellow flowers bloom on a flower cluster, then it's time to harvest. Don't leave it too long, or the plant will start to taste bitter.

Choy sum can be harvested in three ways:

- **Harvest the flower buds** – Lots of people love eating the buds, so they focus on growing the plant for its flowers. Harvest the flower stalks as soon as the first few yellow flowers appear, leaving a couple of baby flower shoots to continue to grow for your next harvest. Keep doing this until the plant loses its vigour, then plant a fresh batch of seeds.
- **Harvest the whole plant** – Allow your choy sum to grow until the first yellow flowers appear, then cut off the entire plant at the base and eat everything in one go.
- **'Cut and come again'** – You can treat choy sum as a regular green and harvest the leaves. Keep doing this until the plant starts to flower, then harvest the entire plant.

## STORING

Store choy sum unwashed in a re-usable container in the fridge for up to a week.

## COOKING

You can use choy sum as a substitute for bok choy – chop it up for fried rice, or add it to curries or stir-fries. Choy sum is tender and it cooks quickly, so add it towards the end of the cooking process.

### Sizzling garlic choy sum

While choy sum is a versatile ingredient, here's a simple way to cook it that celebrates the freshness of homegrown harvests.

Cut the choy sum in half. Finely chop the garlic cloves.

Bring a saucepan of water to a boil. Add the choy sum, and cook for 2 minutes or until the stems become vibrant green.

Transfer the choy sum to a bowl, and top with garlic.

Drain the water from the saucepan. Heat the vegetable oil in the saucepan over medium heat until it starts to bubble. Pour the oil on top of the choy sum and garlic. Serve warm.

**SERVES 2–4**

250 g (9 oz) choy sum

2 garlic cloves

1 tablespoon vegetable oil

# MUNG BEAN SPROUTS

*Vigna radiata*
Mandarin: 绿豆芽 – lǜ dòu yá
bean sprouts • green gram • kacang hijau • monggo

| | PLANT TYPE | Indoor |
|---|---|---|
| | POT FRIENDLY? | Yes, *jar* friendly |
| | SUN | No sun – darkness please! |
| | WATER | Rinse 2–3 times a day |
| | FOOD | None |

Magnificent mung bean sprouts are a popular ingredient in Asian cuisines. They have a light and sweet taste, and add a crunchy texture to stir-fries, noodles and soups. The nutrient-dense sprouts are packed with vitamins and minerals.

Mung bean sprouts have a short shelf life, so sprouting your own means that you can have them fresh whenever you need them. Luckily, they're ridiculously easy to grow! They don't require soil or light, can be grown indoors all year round, and take only a few days before they're ready to eat.

*When rinsing your sprouts, save the used water for your plants. Also, soy beans can be sprouted in the same way as mung beans!*

# GROWING

You'll need:

1. **Dried mung beans** – These are most economical when bought in bulk from the supermarket.
2. **A 500 millilitre (17 fluid ounce) glass jar** – This will yield approximately 125 grams (4½ ounces) of mung bean sprouts, which is the perfect portion to use as an ingredient in stir-fries, noodle dishes, and so on.
3. **A mesh lid for your jar** – You can buy mesh lids made for sprouting, or use a cheesecloth and rubber band.
4. **A thick, dark-coloured kitchen towel** – This will stop light from getting to the mung beans, and help them to grow creamy yellow and taste sweet.

**Method** Add 1 tablespoon of dried mung beans to your jar, and rinse them in water, removing any broken ones. This will help to reduce the risk of mould, which can spoil the entire jar of beans.

Once rinsed and drained, soak the beans in warm water in the jar overnight to break their dormancy. The next morning, drain the water from the jar, and rinse and drain your beans twice more.

Turn your jar upside down at a 45-degree angle, and place it over a bowl to catch the excess water. (The angle helps to keep everything ventilated.) Pop a kitchen towel over the jar, so the beans are not exposed to light.

Rinse and drain the jar two or three times a day until the beans sprout (this will take three to five days), and continue to cover the jar with the kitchen towel. Rinsing the beans:

- keeps them clean
- reduces the risk of mould
- improves germination rates
- ensures they're kept moist.

During the process, remove any sprouts that look damaged or odd. If at any time during the process things start to feel slimy or look mouldy, discard everything, sterilise the jar and start again.

# HARVESTING

Once the sprouts grow to your desired size (they will be smaller than supermarket sprouts, as homegrown ones don't grow in *absolutely perfect* conditions), then it's time to clean them. The green bean shells are edible, but if you want to remove them, then pop the sprouts into a bowl of water and swish everything around. The shells will float to the surface, where you can skim them off.

# STORING

Mung bean sprouts can be stored in an airtight container in the fridge for a day or two but should be eaten as soon as possible. It's best to sprout small quantities and eat them quickly rather than sprout big batches and store them. To check the freshness of a mung bean sprout, simply snap it in half. If it snaps easily, then it's fresh. If it's bendy, slimy or mushy, then compost it.

# COOKING

I recommend boiling mung bean sprouts before using them in dishes (even salads), to ensure that they don't harbour any harmful bacteria.

*If you have leftover dried kombu seaweed and you're wondering what to do with it, then try adding it to soups, hotpot (see page 101), fried noodles or even coriander, tofu and sprout salad (see page 219).*

## Simple sesame sprouts

This recipe is the epitome of simplicity without sacrificing flavour. It's our go-to sprouts recipe and exactly the kind of cooking I love to do – not much! Serve it as a veggie side with rice. Scale this recipe up or down depending on how many mung bean sprouts you have.

Chop the chilli (if using).

Bring a saucepan of water to a boil. Add the mung bean sprouts, and cook for 30–60 seconds. Drain the sprouts in a colander, and set them aside for 1 minute to drain further. Transfer the sprouts to a bowl.

Add the sesame oil, light soy sauce, cooking salt and chilli (if using). Mix well, and serve warm.

**SERVES 2–4**

1 fresh chilli (optional)

250 g (9 oz) mung bean sprouts

2 teaspoons sesame oil

2 teaspoons light soy sauce

¼ teaspoon cooking salt

## Pickled bean sprout and seaweed salad

This sweet and tangy salad combines mung bean sprouts with dried kombu seaweed, a pantry staple, and five spice tofu, which I always pick up when I'm at the Asian grocer. If coriander (cilantro) is in season when you're making this recipe, then you can add a bit of it to the salad for a herby touch. The dried kombu seaweed needs to be rehydrated beforehand, so do make time for that.

Rehydrate the dried kombu seaweed by soaking it in a bowl of water for 1–2 hours. Slice it into thin pieces, in a similar shape to the mung bean sprouts. Slice the tofu in similar thin pieces. Finely chop the garlic cloves. Chop the chilli (if using).

Bring a saucepan of water to a boil. Add the sliced tofu, and remove after 1 minute. This is just to slightly soften the tofu.

Add the mung bean sprouts to the same water, and remove after 30–60 seconds. Add the kombu seaweed to the water, and bring back to a boil. Once boiling, cook the seaweed for 1–2 minutes, then remove from the water.

Drain the tofu, mung bean sprouts and kombu seaweed. Place them into a bowl with the garlic, chilli (if using), light soy sauce, vinegar, sugar, sesame oil and ground white pepper. Mix well before serving.

**SERVES 2–4**

15 g (½ oz) dried kombu seaweed

70–80 g (2½–3 oz) five spice tofu (about 2 squares)

2 garlic cloves

1 fresh chilli (optional)

150 g (5¼ oz) mung bean sprouts

1 tablespoon light soy sauce

1 tablespoon white or black vinegar

1 teaspoon sugar

1 teaspoon sesame oil

Ground white pepper, to taste

# OYSTER MUSHROOM

*Pleurotus ostreatus*
Mandarin: 平菇 – píng gū
hiratake • oyster fungus • pearl oyster

| | | |
|---|---|---|
| **PLANT TYPE** | Indoor | |
| **POT FRIENDLY?** | Cultivated in a grow bag or small tub | |
| **SUN** | Indirect sunlight | |
| **WATER** | Mist 2–3 times a day | |
| **FOOD** | Mushroom substrate | |

Head to an Asian grocer, and you'll find many interesting mushrooms – oyster, shiitake, shimeji, enoki, just to name a few. As a grower of edible things, I've always been curious about cultivating my own mushrooms at home.

As it turns out, the oyster mushroom is one of the easiest and fastest growing mushrooms, and there's a variety for every climate. They can also be grown indoors, and they're great for apartments and small spaces.

Oyster mushrooms grow in clumps. They have short stems, delicate gills, a smooth oyster-shaped cap and a slight seafood flavour. I love stir-frying them with leafy greens for a completely homegrown dish. They have a meaty texture and resemble chicken when sliced. I have to look twice when cooking because their appearance is so similar!

*Aside from the right temperature, the top three things you need to provide for your oyster mushrooms (and mushrooms in general) are humidity, fresh airflow and indirect light.*

## VARIETIES

**White**, **blue (blue grey)** and **chocolate** (*Pleurotus ostreatus*) oyster mushrooms fruit when the temperature reaches 10 degrees Celsius (50 degrees Fahrenheit), perfect for growing indoors in cooler climates or during winter. There are also **pink** (*P. djamor*) and **golden** (*P. citrinopileatus*) oyster mushrooms, which fruit when the temperature is at least 18 degrees Celsius (64 degrees Fahrenheit). This makes them more suited to warmer climates or growing through summer.

## GROWING

An easy way to grow your own oyster mushrooms is to start with a mushroom kit from a dedicated mushroom store, garden centre or nursery. A kit consists of a bag of mushroom-growing 'stuff' packaged in a cardboard box or bucket.

The bag of 'stuff' is the growing medium (the substrate) inoculated with mycelium. For oyster mushrooms, this growing medium can be moistened sawdust pellets, straw or coffee grounds. The mycelium is the white, mouldy-looking bits present on the medium. This is the root system of the oyster mushroom that colonises the growing medium, turning it white all over. Once this happens, it's ready to fruit – in other words, grow some beautiful mushrooms!

You'll need to cut an X (cross) into the mushroom kit, so the mushrooms have somewhere to fruit from. Oyster mushrooms are a side-fruiting species, so you'll have to make the cut on the side of your box. It's totally fine (and preferable) to slice into the substrate when making this hole, as 'damaging' the surface can encourage mushrooms to grow.

Oyster mushrooms require both humidity and lots of fresh air to grow. Mist where you cut two or three times a day. The X cut allows in fresh air, while the flaps of the X help to keep the substrate humid.

Oyster mushrooms also need indirect light – place them somewhere you can comfortably read a book. Keep the kit out of direct sunlight. For some oyster mushrooms, the lower the light, the lighter the cap colour. Cooler and warmer temperatures can also change their colour.

**Fruiting** When oyster mushrooms start to appear, they will grow quickly and can double in size every day. Continue misting two or three times a day during fruiting.

*The oyster mushroom gets its botanical name from the shape of its cap.* **Pleurotus** *means 'sideways', while* **ostreatus** *means 'like an oyster'.*

*When your mushroom kit has finished fruiting, pop it into the compost. The mycelium is great for your garden.*

## HARVESTING

Harvest oyster mushrooms right before the mushroom caps curl up. Once the caps turn up, they drop their powdery spores. Some people are allergic to mushroom spores and they're also a pain to clean, so it's best to harvest the mushrooms before this happens. Mushrooms with curled caps can still be eaten, but they're past their best and won't store as long in the fridge.

Harvest the entire clump of mushrooms together by twisting it off at the base. Ensure that all of the stalks are removed, so new mushrooms can eventually grow in their place. Some of the substrate might come away with the mushrooms; this is a good thing, as it means that the space is cleared and ready for the next batch.

**Growing your second (and third!) flush** You can usually get two or three harvests from a mushroom kit, with each subsequent flush having fewer and fewer mushrooms. After each flush, you'll need to let your growing medium rest for one or two weeks before starting the growing process again.

You may need to rehydrate the growing medium after the first or second harvest. If so, pop your mushroom bag, hole down, into a large bowl of water. Place something heavy on top to prevent it from floating, and allow it to rehydrate overnight. Mushrooms are up to 90 per cent water, and that water comes from the growing medium.

## STORING

Store your freshly harvested oyster mushrooms in a paper bag with the top scrunched closed, rather than in an airtight container, as mushrooms breathe and release moisture so need airflow to keep fresh. They can be kept in the fridge this way for up to seven days, if they have been picked before their caps have turned up.

## COOKING

Try oyster mushrooms in any of the recipes that call for mushrooms in this book. Some people prefer to remove the chewy stems of oyster mushrooms before cooking, but I think they taste great and I eat the mushrooms whole. If you remove the stems, then you can throw them straight into the compost – but why not save them to make veggie stock? Collect mushroom stems and other veggie scraps in a jar in the freezer; once you have enough, boil everything in water to create a delicious stock.

204

### OTHER ASIAN MUSHROOMS TO GROW AT HOME

If you love growing your own oyster mushrooms, then why not give these varieties a go:

**Enoki**
金针菇 • jīn zhēn gū
These mushrooms are a fun challenge, as they require a cool temperature of 10–15 degrees Celsius (50–59 degrees Fahrenheit) to fruit.

**King oyster**
杏鲍菇 • xìng bào gū
These are similar to oyster mushrooms but have large stems and small caps. The fleshy stems are used as a meat substitute.

**Shiitake**
香菇 • xiāng gū
These mushrooms are traditionally grown outdoors on tree logs but are also available in mushroom kits.

*Mushrooms take in oxygen and release carbon dioxide, so they need a continuous supply of fresh air to grow properly.*

# SNOW PEA (MANGE TOUT) SHOOTS

*Lathyrus oleraceus*
Mandarin: 豆苗 – dòu miáo
snow pea sprouts • snow pea tips

| | PLANT TYPE | Indoor |
|---|---|---|
| | POT FRIENDLY? | Yes |
| | SUN | Indirect sun |
| | WATER | Keep moist |
| | FOOD | Light feeder |

Snow pea shoots are a popular veggie side in Chinese restaurants. These young, tender shoots have the sweetness of spring, with a subtle hint of green peas. They are grown indoors, near a window, at any time of the year. I love growing them when there's not much else to harvest, or when it's too cold to grow anything outdoors. All you need is a small container, a little soil and a sunny windowsill. You can harvest in less than two weeks!

*You can also harvest snow pea shoots from the tips of the plants that you're growing for pods. However, if you want to grow a large quantity of shoots, then it's easier to sow seeds in a dedicated container and harvest from there.*

*I find that a surface area of around 240 square centimetres (37 square inches) - about the area of a 13 × 18 centimetre (5 × 7 inch) lunchbox - yields around 70 grams (2½ oz) of snow pea shoots. Use this as a rough guide to calculate how much you can grow in your container.*

# GROWING

You'll need:

- **A shallow container** – Choose something (such as a re-usable lunchbox) that can hold 2–3 centimetres (¾ –1 inch) of growing medium. I like using a glass container, as I can watch the roots grow and better monitor soil moisture.
- **Snow pea seeds** – You can buy seeds for sprouting from garden centres and seed-supply stores. Make sure that you're getting snow pea seeds (*Lathyrus oleraceus*) and not sweet pea seeds (*Lathyrus odoratus*). Sweet peas look like their edible cousins but are grown for their fragrant flowers; in addition, they're toxic and can't be eaten. You can also use dried peas (for eating) bought from the supermarket.
- **Water in a spray bottle** – Have this handy to mist the growing medium.
- **Some soil, potting mix or compost** – Growing shoots is a bit different from growing microgreens. With microgreens, you're only growing plants to the stage of cotyledons and the first set of leaves. Because a plant's seed contains the energy needed to support cotyledons and the first set of leaves, microgreens need the growing medium only for the purposes of providing moisture. Hence, they can be grown with plain water, grow mats or paper towels. When it comes to growing young shoots, you want the plant to grow further than the microgreens stage, and to produce a few more sets of leaves. Because of this, you need to use a proper growing medium, such as soil, potting mix or compost.

**Method** Spread a thick layer of seeds over half of the surface area of your container. Cover with water, and let the seeds soak overnight (or up to 24 hours) to soften the hard outer shell and speed up germination. The seeds will swell and cover the whole surface area of the container.

Remove the seeds temporarily, and add 2–3 centimetres (¾–1 inch) of growing medium to the container. Water just enough to moisten the growing medium. Spread the soaked seeds over the top.

Now, it's time to germinate the seeds. Successful germination requires that the seeds and growing medium stay moist and in darkness. Do this by:

- resting a lid on top of your container, to help reduce water evaporation
- covering the container with dark-coloured kitchen towel
- misting the seeds with water if they start to dry out.

Your snow pea seeds should germinate in around two or three days. Once this happens, remove the kitchen towel and move your container to a windowsill with indirect sunlight. Check on your container daily, and keep the seedlings and growing medium consistently moist (but not flooded) by gently watering.

## HARVESTING

Your snow pea shoots will be ready to harvest after ten to fourteen days. Harvest the shoots when they're around 10–15 centimetres (4–6 inches) tall by snipping across the base of your plants around 2 centimetres (¾ inch) from the surface of the growing medium. Pop the growing medium into the compost or garden and start again. If you'd like to have a continual harvest, make it part of your routine to start a new tray every weekend.

## STORING

Store your harvested snow pea shoots in an airtight container in the fridge. They should keep for up to a week but are best used fresh.

## COOKING

Tender as can be, snow pea shoots remind us that veggies taste incredible when harvested young. The best part about cooking with homegrown shoots is that you don't need to wash or chop them, and they cook in less than a minute.

### Easy pea-sy stir-fry

Welcome to the most delicious greens you'll ever eat! This garlicky dish with a subtle pea flavour has only four ingredients. Could it be any simpler? Scale up the recipe according to how many shoots you grow.

Chop the garlic.

Heat the vegetable oil in a small sauté pan over medium heat. Add the garlic, and cook until fragrant.

Add the snow pea shoots and cooking salt to taste, and stir-fry for 1 minute. Serve warm.

208

---

**MORE IDEAS FOR USING SNOW PEA SHOOTS**

- Use them in stir-fries, but remember to add them last because they cook quickly.
- Sprinkle them on sandwiches.
- Pop them on top of eggs on toast – the delicate shoots will really elevate your breakfast at home!
- Use them as greens in garden salads.

*Broad bean shoots are also tasty and can be grown in the same way as snow pea shoots. You can use dried broad beans bought from the supermarket as the seeds*

**SERVES 1**

1 garlic clove

2 teaspoons vegetable oil

70 g (2½ oz) snow pea (mange tout) shoots

Cooking salt, to taste

# HERBS

# CHRYSANTHEMUM FLOWERS

*Chrysanthemum indicum; C. morifolium*
Mandarin: 菊花 – jú huā
Chinese chrysanthemum • wild chrysanthemum • yeju hua

| | | |
|---|---|---|
| | PLANT TYPE | Perennial |
| | PLANT FAMILY | Daisy family (Asteraceae) |
| | PLANT SIZE | Medium bush to 60 cm (24 in) tall |
| | POT FRIENDLY? | Yes, 30 cm (12 in) or larger |
| | SUN | Full sun or part-sun |
| | FROST | Frost hardy |
| | WATER | Regular watering |
| | FOOD | Medium feeder |

Chrysanthemum tea is a popular summertime drink in China, Japan and South Korea. The small, daisy-like flowers impart a sweet, earthy, floral flavour. They are usually used whole and retain their delicate shape in the cup. Chrysanthemum tea has a calming effect, so I love brewing it in the late afternoon and evening before bed. In addition, chrysanthemum is a Chinese medicine plant traditionally used to treat colds, migraines and inflammation.

Some people liken chrysanthemum to chamomile; however, while both are floral, the flavours do differ. I highly recommend giving it a go, whether or not you enjoy chamomile tea.

*Chrysanthemums are pollinator-friendly plants, as bees absolutely adore the flowers.*

*When growing chrysanthemum flowers for tea, stick to the species listed in the Varieties section above. While the flowers of all chrysanthemum species are edible, not all of them taste good. I've tried a couple of other chrysanthemums in my garden that I definitely don't rate!*

## VARIETIES

*Chrysanthemum indicum* and *C. morifolium* are the two species most often used for tea. Both yellow and white flowers are available, with the yellow flowers being more common. In Asia, other chrysanthemum species are also used to make tea; however, the specifics of those are unclear.

## GROWING

You need only one plant in a garden, as it will grow into a bush and produce many flowers. Start with a seedling, and plant it at the beginning of spring in a full-sun or part-sun position. Keep it well-watered, especially during warm weather. Chrysanthemums are quick growing and will bloom from late summer to autumn. In cool climates, they become dormant in winter and sprout again in spring.

Chrysanthemums are fuss-free plants. The ones in my garden grow without much care, which is why I love them so much. They are also pollinator-friendly plants, and bees adore the flowers.

**Propagation** Chrysanthemums can be propagated in a few different ways. The easiest is to bend some of the lower stems of the bush so they rest on top of the soil – you can bury them just under the surface if you like. The stems will develop roots, and then you can snip the stems from the mother plant and pot them up to make new plants.

## HARVESTING

Pick chrysanthemum flowers when they're two-thirds open, ideally in the morning – that's when they're at their best. If you want to dry chrysanthemum flowers for later use, pop them in a single layer on a tray, in indirect sun. It's best to dry them indoors, so they don't fly away in the wind. Turn them occasionally so they dry evenly. You can also use a dehydrator or oven on the lowest temperature setting, but I find that they dry fine without one.

## STORING

Once dried, store the chrysanthemum flowers in an airtight jar out of direct sunlight in the pantry. They will easily last for up to a year – until your next harvest.

## COOKING

To make your own chrysanthemum tea, follow these step-by-step instructions:
1. Rinse eight to ten chrysanthemum flowers.
2. Pop the flowers into 1 cup (250 millilitres) hot water.
3. Steep the flowers for four to five minutes.
4. Serve the tea hot or with ice.
The flowers can be re-used throughout the day until they lose their flavour. And they look especially gorgeous in a glass teacup!

# CORIANDER (CILANTRO)

*Coriandrum sativum*
Mandarin: 香菜 – xiāng cài
Chinese parsley

| | | |
|---|---|---|
| 🔖 | **PLANT TYPE** | Cool-season annual |
| ⬮ | **PLANT FAMILY** | Carrot family (Apiaceae) |
| ⤢ | **PLANT SIZE** | Small, around 40 cm (16 in) tall, but when it flowers it'll grow up to 70 cm (28 in) tall |
| 🪴 | **POT FRIENDLY?** | Yes, plant one clump in a 20–30 cm (8–12 in) pot or a few in a larger pot |
| ✳ | **SUN** | Full sun or part-sun |
| ❄ | **FROST** | Frost tolerant |
| 💧 | **WATER** | Regular watering |
| ⁂ | **FOOD** | Light feeder |

I love coriander and can't get enough of it, but it certainly can be a divisive herb. Some adore the citrusy flavour it adds to dishes, while others think it tastes like soap. This is because of a special gene that some people possess!

Homegrown coriander is so much fresher and more vibrant than the bunches you buy from the supermarket. I've been able to store freshly picked coriander in the fridge for two weeks! Store-bought bunches, on the other hand, always seem to go limp and slimy after just a couple of days.

Being a small, fast-growing plant, coriander is relatively easy to grow and largely free from pests and diseases. It can be frozen and used throughout the year.

# GROWING

Coriander is best started in autumn, so direct sow the seeds when the weather begins to cool. Don't sow seeds in spring or summer, as the plant has a tendency to bolt before you get a chance to harvest.

Interestingly, coriander seeds are polyembryonic – each is actually a pair of seeds. This is nature's way of suggesting that these plants are best planted in groups to allow them to provide support for each other. So, sow a few seeds together, and allow them to grow into a bushy clump.

Coriander seeds are best direct sown where you want them to grow. Sow groups of seeds together on the surface of the soil, around 30 centimetres (12 inches) apart. Cover the seeds with a thin layer of vermiculite (which also helps you to see where you sowed them in the garden).

You can sow seeds in punnets in the same way; however, be sure to transplant the seedlings while they're still young.

It's best not to transplant mature coriander plants, as they grow a long taproot that looks like a small carrot, and this is easily damaged.

Coriander plants are fond of bolting – it's pretty much the only challenge with growing them, as everything else is so easy. Naturally, we want our plants to grow for as long as possible before bolting, so we can have a longer harvest time. Here are my top tips on how to convince your coriander plant not to bolt so early:

- **Look for slow-bolting varieties** – These have been cultivated to be more heat tolerant and bolt less quickly than standard varieties. Slow-bolting coriander will still bolt, just a little later than other varieties.
- **Plant in a shady spot** – If you live in a warm climate, then plant coriander where it will receive lots of shade (for example, in the shadow of taller winter veggies, such as chrysanthemum greens). This will protect it from too much warmth and sunshine, which can contribute to early bolting.
- **Succession plant** – Sow a new batch every month, so that you always have coriander to harvest, even if older plants start to bolt.
- **Keep a consistent watering schedule** – Coriander plants require lots of water when they're young and still establishing themselves. Dry spells and inconsistent watering can cause plants to bolt.

# HARVESTING

Harvest coriander as a 'cut and come again' green. Once your plant has established itself and reached 30–40 centimetres (12–16 inches) in height, simply snip off a portion of the stems at the base whenever you need them.

**Keeping the seeds** After coriander plants bloom, and the flowers start to die, little seeds will form in their place. Let the seeds begin to turn brown on the plant, then cut off the flower stems and pop them upside down in a paper bag to collect the seeds. The dried seeds can be used as a spice in the kitchen – just store them in an airtight jar like other spice, and keep the jar in the pantry.

If you want to sow the seeds the following season, then always collect them from the plant that bolts last. Don't collect from plants that bolt first, as you are selecting for the unwanted early-bolting trait.

# STORING

Stand the freshly harvested coriander leaves in a jar of water in the fridge for up to a week. I also like to chop up excess leaves and freeze them in ice-cube trays for use throughout the year. They're best used in cooking (stir in at the end of the cooking process), as they can become mushy when defrosted.

# COOKING

I like to use coriander as a garnish, in the same way you'd typically use spring onions (scallions). While spring onions grow all year round, they tend to slow down when the weather is cool – which just happens to be when coriander is growing well. I often garnish my seafood dishes, curries, salads and soups with the frilly leaves of coriander. If you love this herb as much as I do, then you can also use it like you would a veggie in a salad.

216

## THE MAGIC OF COMPANION PLANTING

I like to think of coriander as a two-phase plant, so I can be grateful for and appreciate its short life cycle. The first phase is when it grows delicious herby greens for me. The second phase is when it bolts, and I leave the flowers as a gift for the insects that work so hard to help me in the garden.

Coriander makes a great companion plant, as its flowers attract beneficial insects and pollinators. In fact, if you're a coriander-tastes-like-soap person, you could still plant it in the garden and treat it as a flower instead of a herb.

# Ginger soy barramundi with coriander

I love making steamed barramundi when coriander (cilantro) is in season. This recipe is simple but packed full of flavour. It's based on a 300 gram (10½ ounce) barramundi fillet, which is best steamed for 8 minutes. If you have a larger fillet - perhaps 400-500 grams (14-18 ounces) - then steam it for 12 minutes. To give the barramundi some oomph, you can drizzle 1 teaspoon light soy sauce on top of the fillet before serving. This dish goes well with rice and pickled veggies (see page 72).

Remove the coriander leaves from the stems; you can chop the leaves or leave them whole.

Rub the cooking salt, ground white pepper and Chinese cooking wine onto both sides of the barramundi fillet. Place the ginger on top of the fillet. Set the fillet aside in a shallow, heatproof bowl (I use a ceramic one) for 10 minutes to marinate.

Bring water to a boil in a steamer. Once boiling, reduce the heat to medium. Transfer the fillet (still in its shallow bowl) into the steamer, and steam for 8 minutes.

Turn off the heat, and let the fillet sit in the steamer for a further 3 minutes. Remove the fillet from the steamer, top with coriander leaves, and serve hot.

**SERVES 2**

5–10 sprigs of coriander (cilantro)

½ teaspoon cooking salt

1 pinch of ground white pepper

1 teaspoon Chinese cooking wine

300 g (10½ oz) barramundi fillet

1 teaspoon finely sliced ginger

# Coriander, tofu and sprout salad

When you've harvested a whole heap of coriander (cilantro), make this salad as a side dish. Its light and refreshing flavour is always a welcome addition to our winter meals.

Chop the coriander into 5 centimetre (2 inch) lengths. Thinly slice the five spice tofu, and finely chop the garlic cloves.

Bring a saucepan of water to a boil. Add the mung bean sprouts and cook for 30–60 seconds, then remove. Place the sprouts into a strainer, and set aside to strain and cool for a few minutes.

Place all of the ingredients into a large bowl, and mix to combine. Transfer to a plate, and serve cold.

**SERVES 4**

50 g (1¾ oz) coriander (cilantro)

100 g (3½ oz) five spice tofu

2 large garlic cloves

250 g (9 oz) mung bean sprouts

3 teaspoons light soy sauce

3 teaspoons white vinegar

2 teaspoons sesame oil

1 teaspoon sugar

½ teaspoon sea salt

# CURRY LEAF TREE

*Murraya koenigii*
Mandarin: 咖喱叶 – gā lí yè
curry patta • kadi patta • sweet neem

| | | |
|---|---|---|
| **PLANT TYPE** | Perennial | |
| **PLANT FAMILY** | Rue family (Rutaceae) | |
| **PLANT SIZE** | Large tree to 5 m (16 ft) tall but can easily be kept small (to 1 m/3 ft) in a pot | |
| **POT FRIENDLY?** | Yes, 30 cm (12 in) or larger | |
| **SUN** | Full sun or part-sun | |
| **FROST** | Frost sensitive | |
| **WATER** | Regular watering | |
| **FOOD** | Light feeder | |

The curry leaf is an aromatic herb essential in Indian (especially South Indian) and Sri Lankan cooking. The small, tender leaves are infused in oil at the start of the cooking process to add a unique citrusy and woody flavour to curries and other dishes. I first tried fresh curry leaves in a home-cooked curry from our neighbours who grow their own curry leaf tree. The leaves added such a subtle but necessary touch to the dish that I was inspired to grow my own curry leaf tree to make better curries.

The curry leaf tree is a small, upright tree native to India and Sri Lanka. It has delicate, drooping leaves and will happily live in a pot all its life.

Despite the name, curry leaves are not the fresh version of curry powder. The two are entirely different things and not used interchangeably. Curry leaf is a herb, whereas curry powder is an Indian spice mix and usually doesn't contain any curry leaf at all. Both, however, are added to curries.

# GROWING

Head to your local nursery to pick up a baby curry leaf tree, but make sure that you buy the right plant. The curry leaf tree (*Murraya koenigii*) is easily confused with the curry plant (*Helichrysum italicum*). While the names are easy to mix up, their looks are not.

The curry leaf tree looks like a tree, with large leaves made up of many smaller leaflets. On the other hand, the curry plant is a small, silver-leafed shrub that looks similar to a rosemary bush. While it also has a strong curry-like aroma, it is not the same plant that's used in cooking.

The best time to transplant your baby curry leaf tree is after the last frost in spring. This will allow it to establish through summer before cold weather hits again.

A happy curry leaf tree in the ground can grow up to 5 metres (16 feet) in height. That is much taller than what you can easily reach, and the tree will produce more leaves than one household could ever use. Mature trees send out vigorous root suckers that can pop up all over the garden. Further, the berries that grow on large trees are difficult to manage, and birds can drop the seeds in unwanted places, making the plant invasive.

Hence, it's best to keep your curry leaf tree well-pruned to your desired height, or, better yet, grow your curry leaf tree in a pot. The pot will help to contain its size, while still providing plenty of leaves to harvest.

It's important to protect curry leaf trees from the cold and frost. If you live in a cooler climate, then growing your tree in a pot will mean that you can move it indoors or to a sheltered position for protection when the weather cools down. Although the trees are usually evergreen, in cooler climates they become dormant in winter and shed their leaves (which grow back in spring as the weather warms up).

Curry leaf trees prefer full sun in cooler climates. A position in part-sun in warmer climates will protect the leaves from sunburn.

**Pruning** Your baby curry leaf tree will likely start off growing as a single upright stem. Snip off the growing tip when it reaches the height at which you'd like it to branch. This will trigger the plant to pop out two side branches, which leads to a bushier plant and more curry leaves for you!

## HARVESTING

Harvest fresh leaves whenever you need them! Do this by cutting off the entire leaf at the base of the stem. Or harvest a bunch of leaves by snipping off the growing tips; this has the bonus of encouraging the plant to branch out and grow bushier.

## STORING

If you've harvested a large batch of curry leaves – say you were pruning your plant and cut off lots of leaves – then the best way to store them and retain their flavour is to freeze the leaves whole. Frozen leaves will keep their flavour for 6–12 months, until your curry leaf tree springs back into action.

Another option is to air-dry your curry leaves on a tray; however, this doesn't keep their flavour as well as freezing them.

## COOKING

Curry leaves are added at the start of the cooking process. Sauté them in oil to release the aromatic flavour, and then continue with the rest of your cooking. The leaves are tender enough to leave in and eat, or you can take them out before serving – the choice is yours!

*Curry leaf trees can start to flower and fruit after their second year.*

*Little white flowers will appear, followed by black berries, which contain a seed.*

*If you don't want to collect the seeds for germination, then remove the berries when they form.*

*This will also stop birds from eating the berries and dropping seeds in unwanted places.*

# GARLIC

*Allium sativum*
Mandarin: 大蒜 – dà suàn
ayo • ayuu • rocambole • stinking rose

| | | |
|---|---|---|
| 🌱 | **PLANT TYPE** | Perennial; usually grown as an annual |
| 🌰 | **PLANT FAMILY** | Onion family (Alliaceae) |
| ⤢ | **PLANT SIZE** | Upright plant, 40–60 cm (16–24 in) tall, 10–15 cm (4–6 in) wide |
| 🪴 | **POT FRIENDLY?** | Yes, grow 6–8 cloves in a 50 cm (20 in) pot |
| ☀ | **SUN** | Full sun |
| ❄ | **FROST** | Frost hardy |
| 💧 | **WATER** | Regular watering |
| ⚬ | **FOOD** | Heavy feeder |

Garlic is such a joy to grow at home. You can experiment with hundreds of different varieties, compared to the few you can buy at the supermarket. Some are hot and pungent, while others are gentle and sweet, and have a rich caramel flavour when roasted. Some have large cloves and others have small cloves. Some store for a few months, whereas others keep for a year before they sprout. But most of all, homegrown garlic is fresh, crisp and bursting with flavour.

Garlic requires cool weather to properly form bulbs, so it's best suited to places with temperate and cool climates; however, some varieties can be grown in a subtropical climate.

*The strong scent of garlic plants makes them great companion plants, as it helps to ward off pests in the veggie patch.*

# VARIETIES

Garlic can be divided into hard-neck and soft-neck types.

| HARD-NECK GARLIC | SOFT-NECK GARLIC |
|---|---|
| • Grows scapes (stiff, central flower stems that need to be pruned off to grow full-sized bulbs); scapes can be used in cooking | • Does not grow scapes |
| • Scapes leave a hard neck, making these varieties difficult to braid for storage purposes | • Without scapes, the neck of the garlic is soft and easy to braid for storage purposes |

Many different varieties exist within these groups, each with their own flavour, size, colour, growing time, preferred climate and storage length.

Select garlic varieties that suit your climate. If you live in a cool climate, then you'll have a large selection from which to choose. If you live in a warm or subtropical climate, then be sure to select a subtropical day-length-neutral variety. Unfortunately, gardeners in tropical regions will have to give this plant a miss, as garlic needs a cold winter to grow properly.

After that, you can look at other characteristics. I usually choose varieties based on storage length and clove size. I prefer long-keeping garlic (such as the **Silverskin Group**), as I don't have to preserve it, and I find that large cloves (such as those of the **Porcelain Group**) are much easier to use than small ones (such as those of the **Artichoke Group**). I'm not too fussed with the taste, as all homegrown garlic has a rich and complex flavour.

# GROWING

Garlic is usually planted in autumn and harvested in spring and summer. Check with local gardeners to find out the best time to plant in your area.

To grow garlic, you'll need to get your hands on **seed garlic**, as these bulbs are sold specially for planting. They differ from regular supermarket garlic in that they're:
- grown locally and better suited to the climate
- not sprayed with growth inhibitors, and will sprout when planted (sometimes supermarket garlic is sprayed to prevent it from sprouting too early)

- free of diseases
- large, which leads to large bulbs
- named varieties, so you know the exact characteristics of the garlic you're growing.

You'll only need to buy seed garlic once, as you can save your own after your first season to replant.

Keep your seed garlic whole until it's time to plant. Then separate the cloves from the bulb, leaving the papery skin on, and sort them into two piles – big cloves and small cloves. Plant only the big cloves, as they produce bigger garlic bulbs. Smaller cloves can be used in the kitchen – consider this an early taste test of the garlic you're about to grow.

Soak your large garlic cloves in diluted seaweed tonic overnight before planting. This helps roots to develop more quickly. Plant your soaked cloves directly into the garden, at a depth of 5 centimetres (2 inches). Space them approximately 15 centimetres (6 inches) apart; the more space you give garlic, the bigger the bulbs will be.

Choose a full-sun position with well-draining soil. Garlic needs good drainage, as it'll rot in waterlogged soil. If you're growing it in a pot, don't use a self-watering pot – a regular pot is best.

As garlic takes a long time to mature – at least six months (and up to eight months where I live) – make sure that the spaces where you plant the cloves aren't double-booked come spring and summer. Somewhere at the edge of a garden bed is a good idea, although ensure that they're getting enough sun and not being shaded by other plants.

Throughout the growing season, be sure to keep the growing area well weeded. Garlic dislikes competition from weeds.

Garlic plants spend the first part of their life growing their leaves, and the second part growing their bulbs. It's only in the last few weeks before harvest that garlic bulbs put on most of their size. So, if you're having a dig around in winter to check the progress of your garlic, don't be disappointed if you see nothing there – bulking up is the last thing on their to-do list.

226

## THE GREAT E-SCAPE

If you're growing a hard-neck garlic variety, then prune off any scapes (stiff central flower stems) as soon as they form in order to get big garlic bulbs. Garlic scapes are common in Asian cooking and are often sold in bunches at Asian grocers. You can chop up your pruned scapes and use them in place of spring onion (scallion), garlic chives or garlic in dishes – try it in my garlic chive egg scramble recipe on page 110. They can also be roasted or used in stir-fries.

# HARVESTING

Come harvest time, the leaves will start to turn yellow. Once this happens, stop watering your garlic plants to allow the soil to dry up. When only five green leaves remain, dig around one of the bulbs to see if it looks large and ready to harvest. If it is, then you'll know that all of the other bulbs of the same variety are ready.

Don't wait until all the foliage dies, as by this time it will be too late. The bulb's papery skin will have started to break down, and the cloves will have begun to separate from the bulb, causing them to rot.

Harvest your garlic after a period of dry weather, as it's no fun digging up mud-covered bulbs! Use a small garden fork to loosen the soil around the bulbs and to gently lift them out.

**Curing the bulbs** If you don't plan to eat all your garlic immediately, then cure the bulbs. Gently brush any excess dirt off the bulbs. Place them on an oven tray and position them under shelter, in a well-ventilated area without direct sunlight. I like to dry mine indoors with the door open, or under the carport.

Leave the bulbs for three or four weeks, and rotate them so they dry evenly. During this time, the outer skins will dry, forming a protective layer over the cloves.

# STORING

Once the bulbs are cured, trim the leaves and roots, leaving just a little of each attached. (If you grew soft-neck garlic, then you can braid the leaves rather than cutting them off.) Set aside your biggest and best bulbs to replant next year, and store all of the bulbs in open paper bags at room temperature out of direct sunlight. How long your garlic stores for will depend on the variety.

A good way to manage your harvest so you can cook with homegrown garlic for an entire year is to freeze a portion of your bulbs for use after your fresh ones have run out. I prefer to eat my medium- and long-keeping garlic fresh; I freeze my short-keeping garlic in airtight containers and use it later.

# COOKING

Garlic is an essential ingredient in many Asian cuisines (and cuisines from around the world). Almost every stir-fry dish I cook starts off with a generous spoonful of garlic. Aside from the bulb and scape, other parts of the plant can be eaten as well. In spring, flowers and leaves can be used to add a garlicky flavour to dishes, although allowing flowers to grow or harvesting too many leaves may have an impact on bulb growth.

### HOW TO SUSTAIN YOUR GARLIC HARVESTS

One afternoon, I did the maths on how to sustain a continual garlic harvest year after year.

I calculated that, on average, there are five large cloves per garlic bulb that are worthy of being planted, meaning you should save 20 per cent of what you harvest to grow the same amount each year.

# GINGER, TURMERIC AND GALANGAL

| | | |
|---|---|---|
| PLANT TYPE | Perennials |
| PLANT FAMILY | Ginger family (Zingiberaceae) |
| PLANT SIZE | 0.5–2 m (1½–6 ft), depending on species |
| POT FRIENDLY? | Yes, and recommended; 30–50 cm (12–20 in) pot |
| SUN | Full sun or part-sun |
| FROST | Frost sensitive |
| WATER | Keep soil consistently damp, but don't let plants sit in water |
| FOOD | Heavy feeders |

Ginger, turmeric and galangal are native to Southeast Asia and grown for their aromatic underground rhizomes. All three belong to the ginger family (Zingiberaceae) and are grown in a similar way, so I've grouped them together to make things easier.

Being tropical plants, this trio thrives best in warmer areas. If you live in a cool climate, then you'll need to mimic a tropical environment as best as you can, by growing them during the warm season and keeping them protected through winter.

For cool climates, galangal is the easiest and most cold tolerant of the three, followed by turmeric and then ginger. You'll have to accept smaller harvests, but you'll still get a reasonable yield. Consider it a garden challenge, and see how you grow!

# GINGER
*Zingiber officinale*
Mandarin: 姜 – jiāng
adrack • chiang • inchi • sang keong • shen jiang • ung

Freshly harvested ginger is highly aromatic and juicy. It can be harvested at various stages of growth, depending on how you want to use it. Aside from in cooking, ginger is also fabulous for travel sickness and nausea. Growing to 80 centimetres (30 inches) tall, ginger plants have small leaves on upright stems and look like small bamboo plants.

# TURMERIC
*Curcuma longa*
Mandarin: 姜黄 – jiāng huáng
dilaw • khamin • kunyit • nanwin

Turmeric has an earthy flavour and is commonly used as a spice. It will easily stain anything it touches yellow, so take care when handling it. A good tip I've learned is to soak stains in vinegar and boiling water to get them out. This plant has wider leaves than ginger and galangal, and grows up to 50 centimetres (20 inches) tall.

# GALANGAL
*Alpinia galanga*
Mandarin: 高良姜 – gāo liáng jiāng
galanga • greater galangal • lengkuas • Thai ginger

Galangal's flavour is mild ginger crossed with pine/pepper. There are two varieties of galangal: greater galangal (*Alpinia galanga*) and lesser galangal (*A. officinarum*). Greater galangal is the one commonly used in cooking, and the one I'm referring to here. Lesser galangal is stronger in flavour and more commonly used medicinally. Galangal roots look similar to ginger roots but have a glossier skin. The plant grows up to 2 metres (6 feet) in height.

# GROWING

The easiest way to grow these three plants is to pick up a baby plant from the nursery or start your own from a rhizome (underground stem – the part you eat). To grow from a rhizome, begin in early spring.

1. Grab a fresh, plump rhizome from the supermarket. Organic ones are best, as they are unlikely to have been sprayed with a growth retardant to prevent sprouting.

2. Break the rhizome into a few pieces that are approximately 3–5 centimetres (1–2 inches) in size, making sure that each piece has at least one 'knobbly bit' on it – this is where the plant will sprout from. Set the pieces aside on your kitchen bench in a ventilated area for two or three days so the ends completely dry out.

3. Plant your pieces approximately 5–10 centimetres (2–4 inches) deep in a pot of moist growing medium, with the knobbly bits facing up. Don't worry if you get the placement wrong – the young shoots will eventually find their way to the surface.

4. Cover the pot with a lid that is slightly ajar to help keep the moisture in but provide a bit of ventilation at the same time. Place the pot on a heat mat to help the rhizome pieces sprout. Within two or three weeks, you should see sprouts. Once the sprouts are 3–5 centimetres (1–2 inches) tall, gently transplant them into a bigger pot. If you live in a cool climate, keep the pot indoors or in a warm spot until the weather warms up.

In subtropical and tropical climates, ginger, turmeric and galangal are easy to grow. In cooler climates, however, we'll need to trick the plant into thinking it's living it up in the tropics. This means giving it a warm space, humid air and moist soil. For cooler climates, it's best to grow in pots. This will allow you to easily move the plant to protect it in winter.

Choose a pot that is 30–50 centimetres (12–20 inches) in diameter. If you're growing in a 30 centimetre (12 inch) pot, then only grow one rhizome piece. If you're growing in a 50 centimetre (20 inch) pot, then you can place two or three rhizome pieces in the pot. The more space the rhizomes have, the better the harvest.

In warmer climates, give plants a part-sun spot. In cooler climates, it's best to give them full sun. It's important to keep plants in a sheltered spot away from strong or cold winds, as these can affect rhizome development. Rhizomes prefer to be kept damp throughout the season; however, don't let them sit in water or they'll rot. Regular pots with drainage holes work best – don't use wicking beds or self-watering pots.

*Top: Galangal*
*Bottom: Turmeric*

# COOL-CLIMATE GROWING TIPS

If you want to grow these plants but live in a cool climate, then follow the handy hints below to give yourself the best chance at success.

### EXTEND SHORT GROWING SEASONS

Usually, ginger takes eight to ten months to mature, turmeric takes seven to ten months, and galangal takes ten to twelve months. In a cool climate, they may not have enough time to bulk up their rhizomes. If you were to plant at the start of spring and harvest at the end of summer, then the plants would only have about five months to grow; the harvested rhizomes would be small and not yet fully grown.

To overcome this, forgo harvesting your plants in the first year; allow them to go through two growing seasons (so, two summers) before you harvest. This gives the plants closer to a full ten months of warm-season growing and will result in much larger rhizomes.

### GIVE YOUR PLANTS ENOUGH WARMTH

Try these tips to help create more warmth for your plants:

- Position pots against a sunny brick wall, which can act as a heat bank.
- Prioritise spots that have afternoon sun over morning sun, as afternoon sun is warmer.
- Grow plants in a greenhouse or cold frame if you have one.
- Create a greenhouse effect by placing your plants against a sunny wall, and leaning a large window frame (complete with its glass) against the wall.

### PROTECT YOUR PLANTS IN WINTER

As the weather cools down, your plants will start to die down. Keep your plants safe from frost by overwintering them either indoors or in a greenhouse (in the same way as chilli plants; see page 134).

During the cooler season, your plants will be dormant and won't need feeding or watering. In fact, watering in winter will cause the rhizomes to rot.

## HARVESTING

Harvest ginger and turmeric as the plants start to die down when the weather cools. The easiest ways to harvest are to:

- dig up the whole plant, harvest the majority of the rhizome except for one chunk, then pop that chunk back into the soil
- dig up only what you need, and leave the rest of the rhizome in the soil.

In spring, plants will shoot up again and continue to grow as usual.

Galangal may not die back as readily as ginger or turmeric. Instead, wait until it's been growing for at least a year, then harvest what you need and let the rest continue to grow.

## STORING

My favourite way to store all three of these rhizomes is to freeze them in small, airtight containers. I like to thinly slice them before freezing, so it's easy to break off what I need without defrosting the full harvest.

## COOKING

When cooking with ginger, turmeric and galangal, remember this: the larger the rhizome, the stronger the flavour. I personally don't bother peeling the rhizome before cooking (except when I make pickled ginger, to keep the texture consistent). Simply give it a good wash. If there are any hard, knobbly bits that worry you, then you can carefully cut those off. The rest of the skin can be left as is.

*Large galangal plants can be divided into smaller ones and make lovely garden gifts*

## WHICH GINGER IS BEST FOR PICKLING?

If you want to pickle ginger (see the pickled sushi ginger recipe on page 234), then it's best to use immature, pink-tipped ginger, as it's sweeter, milder and more tender than mature ginger. To harvest young ginger, simply dig it up five to seven months after planting, and look for rhizomes with pink tips. These tips give pickled sushi ginger its signature colour.

### The second-best option

If you miss the immature ginger stage, or you want to make pickled ginger from older store-bought ginger, then look for the youngest, least woody pieces of 'regular' ginger you can find. Older ginger stays yellow when pickled and will have a much stronger flavour.

# Gardener's sambal

Sweet, sour, salty and spicy, sambal is flavourful in every way – and I love adding it to Asian veggie stir-fries (see sambal kang kong on page 183). You can also use it on eggs, in curries, mixed into fried rice, and for so much more. I call this gardener's sambal, as almost all of the ingredients can be grown in your garden (and almost all of them can be found in the pages of this book!).

I prefer to cook with ingredients that are in season at the same time. However, I've made an exception for the lime juice, as it adds a brilliant citrusy tang to the sambal. To ensure that I have citrus juice year-round, I freeze fresh lime (and lemon) juice in ice-cube trays, and defrost when needed. To store the sambal, you'll need a clean and sterilised glass jar with a lid (see page 72).

Place all of the ingredients (except 1 tablespoon water and the vegetable oil) into a blender, and blitz until it reaches your desired texture (sambal is often a bit coarse).

Heat the vegetable oil in a small saucepan over low heat. Add the contents of the blender to the saucepan. As the paste will be thick, you'll find that a lot of it is still left in the blender. Pour the remaining 1 tablespoon water into the blender to loosen the rest of the paste, then transfer this to the saucepan. Cook the paste until it has thickened.

Pour the sambal into the jar, and place the lid on. Store it in the fridge for 2–3 weeks – but it will probably be long gone before then!

**MAKES 200 ML (6¾ FL OZ)**

4 cm (1½ in) piece of turmeric (approx. 15 g/½ oz)

4 cm (1½ in) piece of ginger (approx. 15 g/½ oz)

4 cm (1½ in) piece of galangal (approx. 15 g/½ oz)

1 stick of lemongrass

4 garlic cloves

2–10 fresh chillies, to taste

1 tablespoon lime juice

3 tablespoons fish sauce (or light soy sauce)

1 tablespoon sugar

¼ cup (60 ml) water

2 tablespoons vegetable oil

*I usually use 2 tablespoons of gardener's sambal per dish, so this recipe makes enough to flavour five dishes.*

*If you want to make a big batch of gardener's sambal or save some of it for later, then freeze portions in an ice-cube tray and pop a cube into your stir-fries.*

## Pickled sushi ginger

Also known as gari, Japanese pickled sushi ginger is delicious and crunchy. Spicy ginger meets sweet vinegar in this palate cleanser that balances out the flavours of sushi and sashimi. I also like to enjoy it with poke bowls and salads, as a garnish on stir-fries or as a side dish to rice. This is a quick version that can be stored in the fridge for up to a month. To store the pickled sushi ginger, you'll need a clean and sterilised glass jar with a lid (see page 72).

Wash the ginger, and peel it with a spoon. Cut the ginger into paper-thin slices using a mandoline or vegetable peeler.

Transfer the ginger slices to a bowl. Pour boiling water over the ginger slices, and let them sit for 1 minute before draining. For older ginger, repeat this process 2–3 times. This helps to reduce the spicy flavour a little. Transfer the ginger to the jar.

In a small saucepan over low heat, combine the rice vinegar, sugar and sea salt. Stir until the sugar and sea salt have dissolved. Bring the mixture to a boil, then remove the saucepan from the heat.

Pour the mixture over the ginger, and allow it to cool a little. Place the lid on the jar, and pop the jar in the fridge for at least 1 day before eating.

**MAKES 250 G (9 OZ)**

100 g (3½ oz) young ginger (the youngest you can find)

4 tablespoons rice vinegar

2 tablespoons sugar

½ teaspoon sea salt

234

## Yellow rice

Many Asian cultures have their own version of yellow rice. I enjoy making this version to pair with curries. Once the rice is ready, I like to stir in fresh coriander (cilantro), which is usually ready to harvest at the same time I'm gathering dormant turmeric in winter.

Finely chop the garlic cloves and onion. Chop the coriander (if using). Rinse and drain the rice.

Heat the vegetable oil in a saucepan over high heat, and stir-fry the garlic and onion until fragrant.

Add the rice, turmeric, spices, cooking salt and water. Bring to a boil, then cover with a lid and reduce the heat to low. Simmer for 15 minutes or until most of the water has been absorbed.

Remove the saucepan from the heat, and set it aside for 10–15 minutes. Fluff up the rice, stir in the coriander (if using), and serve warm.

**SERVES 2–3**

2–3 garlic cloves

1 onion

1 sprig of coriander (cilantro) (optional)

1 cup (220 g) jasmine or basmati rice

1 tablespoon vegetable oil

1 tablespoon finely chopped fresh turmeric

Spices of your choice, to taste (such as cardamon)

½ teaspoon cooking salt

440 ml (15 fl oz) water

# LEMONGRASS

*Cymbopogon citratus*
Mandarin: 柠檬草 – níng méng cǎo
citronella grass

| | PLANT TYPE | Perennial |
|---|---|---|
| | PLANT FAMILY | Grass family (Poaceae) |
| | PLANT SIZE | Up to 1 m (3 ft) tall, with an ever-widening clump |
| | POT FRIENDLY? | Yes, and recommended; 30 cm (12 in) pot or larger |
| | SUN | Full sun or part-sun |
| | FROST | Frost sensitive |
| | WATER | Light to regular watering, but do not waterlog |
| | FOOD | Light feeder |

Native to India and Southeast Asia, lemongrass is a subtropical and tropical clumping plant with long, upright leaves. In warmer climates, it grows as an evergreen; in cooler climates, it'll die down in winter and resprout in spring.

The flavourful and tender inner stalk of lemongrass is typically used as an aromatic herb in cooking, while the leaves are brewed to make a light, lemony tea. The aromatic leaves can also be used as a natural mosquito repellent.

Lemongrass is one of the easiest herbs to grow - so much so that I recommend you grow it in a pot so it doesn't take over your whole garden. It thrives on neglect - the only thing you need to do is keep it safe from frost.

*It's best to grow lemongrass in a pot, as it's a vigorous, spreading plant that can become invasive in the right garden conditions.*

# VARIETIES

There are two commonly available species of lemongrass: **Cymbopogon citratus** (West Indian lemongrass) and **C. flexuosus** (East Indian lemongrass). *C. citratus* grows up to 1 metre (3 feet) in height, has thick, juicy stalks and is the one you want to grow for cooking. On the other hand, *C. flexuosus* grows up to 2 metres (6 feet) tall, has thinner stalks and is used for scented products (such as aromatherapy oils).

# GROWING

Lemongrass is rarely grown from seed. Instead, buy a nursery plant, or divide an existing plant into smaller clumps. This is best done in winter when the plant has died down. Simply remove the plant from the pot, then gently separate into smaller clumps and repot.

You can also grow lemongrass from store-bought produce. Cut off most of the green tops, leaving about 10 centimetres (4 inches) remaining. Pop the bottom sections in a glass of room-temperature water on a windowsill, with the white parts submerged. Change the water every few days to keep it fresh. Roots should appear after a couple of weeks, and you'll see new growth on top. Once this happens, you can transplant into a pot.

Lemongrass likes to be kept moist during summer but is drought tolerant once established. It's not particularly fussed with much else and will tolerate all soil types, forgetful watering and minimal feeding. During winter, while the plant is dormant, keep it dry or the roots might rot.

If you live in a cool climate, then you can grow lemongrass by leaving it outdoors during the warmer seasons and moving it to a sheltered spot, indoors or into a greenhouse in winter.

Once the weather starts to cool down, lemongrass leaves will begin to turn brown. When this happens, give the plant an all-over haircut to 10 centimetres (4 inches) from the base. It'll then go dormant until it reshoots in spring.

As your plant grows, it's best to divide it into smaller plants every two or three years. This prevents overcrowding, which can lead to smaller stems. Do this in winter, and share the lemongrass love with friends.

# HARVESTING

Lemongrass can be harvested at any time during the growing season. Regular harvesting is a good thing, as it encourages more leaves to grow.

The leaves can be harvested by removing them at the base of the plant. Take care, as they have razor-sharp edges that will cut your hands and fingers.

Lemongrass stalks can be harvested by breaking them at the base, away from the rest of the clump.

# STORING

Extra lemongrass stalks can be stored in airtight containers in the freezer. Extra leaves can be dried and used for tea.

# COOKING

Before using lemongrass stalks in your cooking, peel off the outer layers to reveal the soft, white centre, then bruise them with the back of the knife to release the flavour.

## Lemongrass and pandan coconut rice

Here's a simple way to enhance your rice, by infusing it with the flavours of your summer garden. This dish can be made on the stove or in your rice cooker. The rice pairs well with chicken, seafood and veggie dishes.

**SERVES 2–3**

1 lemongrass stalk

1 cm (½ in) piece of galangal and/or ginger (optional)

1 pandan leaf

1 cup (220 g) jasmine rice

400 ml (13½ fl oz) coconut milk

200 ml (6¾ fl oz) water

½ teaspoon cooking salt

Remove the outer layers of the lemongrass stalk. Bruise the lemongrass with a knife. Slice the galangal/ginger (if using). Tie the pandan leaf in a knot. Rinse and drain the jasmine rice.

Place the lemongrass stalk, galangal/ginger (if using), pandan leaf, jasmine rice, coconut milk, water and cooking salt into a saucepan. Bring to a boil, then simmer over low heat (with the lid on) for 15 minutes or until the liquid has been absorbed.

Remove the saucepan from the heat, and set it aside with the lid on for a further 10 minutes. Fluff up the rice, and serve warm.

238

### LEMONGRASS TEA

When you're giving your lemongrass plant a trim (such as when you're prepping it for winter), you'll be left with lots of leaves. Instead of throwing these away, dehydrate them to make tea! Lemongrass tea can be enjoyed on its own or combined with ginger, mint or green tea.

Lemongrass leaves don't have a lot of moisture in them, so they can easily be dried without using a dehydrator or oven. Simply cut them into 1 centimetre (½ inch) pieces, lay them on a large platter or oven tray, and dry them in indirect sunlight for one or two days. Once the leaves are crisp, transfer them to an airtight container and store them in the cupboard.

# PANDAN

*Pandanus amaryllifolius*
Mandarin: 班兰 – bān lán
screwpine

| | | |
|---|---|---|
| PLANT TYPE | Perennial | |
| PLANT FAMILY | Screwpine family (Pandanaceae) | |
| PLANT SIZE | 1.5 m (5 ft) tall and wide but will grow smaller in cooler climates and in pots | |
| POT FRIENDLY? | Yes, and recommended for temperate climates, 30 cm (12 in) pot | |
| SUN | Full sun or part-sun | |
| FROST | Frost sensitive | |
| WATER | Keep consistently moist but not sitting in water | |
| FOOD | Heavy feeder | |

Anytime I see something pandan flavoured, you best believe I'm getting myself a bite. Pandan has a taste and aroma that is reminiscent of coconut mixed with vanilla, and it's commonly used in cakes and desserts, as well as rice, meat dishes, jams and drinks. I love the scent of pandan and will always follow my nose whenever I catch a whiff of it.

Pandan is well loved and commonly grown throughout Southeast Asia, where it thrives in its natural tropical habitat. It grows as a shrub with strappy scented leaves that fit beautifully in tropical and coastal gardens, and as a bonus it doubles as an edible perennial.

If you live in a warm or tropical climate, then pandan will be a walk in the park to grow. But if you are in a temperate climate, then you'll need to take a few extra steps. For those in cool climates – I'm sorry, but you might have to give this one a miss!

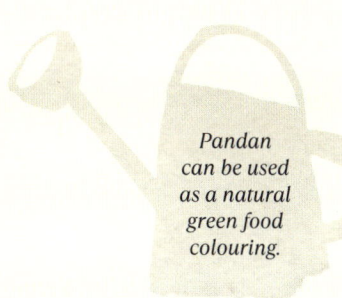

*Pandan can be used as a natural green food colouring.*

# GROWING

Pandan plants are usually available in nurseries during spring. If you know someone with a pandan plant, then you can ask them for a baby plant (if they have one), as happy mother plants shoot out lots of little pups. Once these pups grow two or three aerial roots, they can be snipped off to make new plants.

If you live in a tropical climate, then grow pandan in the ground. For a temperate climate, it's best to start your plant in spring to give it a chance to establish before the weather cools down, and to grow it in a pot. This is because pandan wants to be on a tropical holiday all year round, preferring temperatures of 20–30 degrees Celsius (68–86 degrees Fahrenheit). If overnight temperatures start to hit 10–15 degrees Celsius (50–59 degrees Fahrenheit), then it won't be happy. If you keep it in a pot, then you'll be able to move it indoors in winter.

In temperate climates, you can also consider growing it indoors all year round. When growing indoors, keep it somewhere humid (such as the kitchen or bathroom) and not too close to windows, where it can be quite cold. Alternatively, you can grow it outdoors when the weather is warm enough, and bring it indoors when overnight temperatures start to hit 10–15 degrees Celsius (50–59 degrees Fahrenheit). Take extra care to harden off the plant (see page 50) when you take it outdoors – and reverse the process when you bring it back indoors – as pandan is a sensitive plant in cooler climates.

Pandan loves humidity. Create a humid environment by popping the plant into a greenhouse, placing it in a sheltered area so wind doesn't dry it out, or sitting it on top of a tray containing water and pebbles. Pandan plants like a bit of sun but prefer not to experience super-hot afternoon sun. Keep your pandan plant well-watered but not waterlogged. In winter (while it's indoors), keep the potting mix slightly on the dry side, but maintain the humidity by misting the leaves because indoor heating can completely dry out the plant.

If you successfully keep your pandan plant alive through its first winter, then you've got the hardest bit sorted out. It's only onwards and upwards from here.

# HARVESTING

The oldest leaves are the most flavourful, so harvest from the bottom. Snip off the light-coloured bit closest to the stem of the plant, as this will be bitter. The rest of the leaf is then ready to use in any of your favourite pandan-flavoured dishes.

# STORING

Store pandan leaves on the plant and harvest whenever needed. Alternatively, whole leaves can be stored unwashed in a re-usable container in the fridge for up to a week, or chopped-up leaves can be frozen in an airtight container and kept for up to six months.

# COOKING

Some recipes containing pandan call for pandan juice or extract. This is essentially pandan leaves that have been blended with water (a ratio of about 20 leaves to 1 cup/250 millilitres water) and then strained. I use a nut milk bag, but you can also use cheesecloth. Pandan juice is best used fresh, so make it just before you need it.

If you use pandan leaves whole (such as to infuse rice), then first cut or crush them to release their flavour.

## Pandan mochi

My sister and I love pandan and are always experimenting with ways to include it in desserts. We come back to this pandan mochi recipe again and again because it's so easy to make. If you have leftovers, then store them in an airtight container in the fridge; reheat briefly in the microwave to bring them back to a soft, sticky texture. If you want to make this recipe but don't have pandan leaves, then you can use 1 teaspoon pandan essence (available from Asian grocers) and ¾ cup (180 ml) water instead.

Wash the pandan leaves, and chop them into 2–3 centimetre (¾–1 inch) pieces. Blitz the pandan pieces and the water in a blender. Strain the liquid into a measuring jug using cheesecloth or a nut milk bag, and pour ¾ cup (180 millilitres) of the liquid into a large mixing bowl.

Add the coconut cream, glutinous rice flour and sugar, and mix until it becomes a soft dough. If your dough is too sticky to handle, then mix in a little extra glutinous rice flour.

Using a spoon, scoop out approximately 2 teaspoons' worth of the mixture, roll it into a bite-sized ball, and place it on a plate. Repeat for the rest of the mixture, but make sure that the mochi don't touch or they'll stick together. Dust your hands with extra glutinous rice flour if the dough is too sticky.

Bring a large saucepan of water to a boil, and drop in the mochi one at a time. Stir to prevent the mochi from sticking to the bottom. You might need to make a couple of batches, depending on the size of your saucepan. Cook the mochi for 5–6 minutes or until the colour is uniform and they start to float. They will expand and flatten a little while cooking.

Use a slotted spoon to remove the mochi from the saucepan, and transfer them to a bowl of shredded or desiccated coconut. Roll the mochi balls in coconut while they're hot.

Let the mochi sit until they're cool enough to handle, then serve. I like to start munching on them while they're still warm.

**MAKES APPROX. 30 BALLS**

80 g (3 oz) pandan leaves (about 15–20 leaves)

¾ cup (180 ml) water

⅓ cup (80 g) coconut cream

2 cups (280 g) glutinous rice flour

80 g (3 oz) sugar

1½ cups (110 g) shredded or desiccated coconut, for coating

*Pandan smells so good that it can be used as an air freshener. Pop a few leaves around your home for a delightful tropical scent.*

# SHISO

*Perilla frutescens*
Mandarin: 紫苏 – zǐ sū
beefsteak plant • Chinese basil • Japanese basil • perilla

244

| | | |
|---|---|---|
| | **PLANT TYPE** | Perennial; grown as an annual in cool climates |
| | **PLANT FAMILY** | Mint family (Lamiaceae) |
| | **PLANT SIZE** | 45–80 cm (18–32 in) tall, 40 cm (16 in) wide |
| | **POT FRIENDLY?** | Yes, plant one in a 20 cm (8 in) pot |
| | **SUN** | Full sun or part-sun |
| | **FROST** | Frost sensitive |
| | **WATER** | Regular watering |
| | **FOOD** | Light feeder |

A popular herb in Japanese cooking, shiso has soft, feathery leaves and a delicate flavour with a minty, citrusy kick. Shiso is commonly used as a garnish in sushi and sashimi dishes – those plastic grass things you often see are meant to be fake shiso leaves! – or covered in batter and served as tempura.

Shiso is a member of the mint family (Lamiaceae), but it isn't invasive and weedy like most other mints. Instead, it grows as a small bush like basil. It can be difficult to find in the shops but is easy to grow, especially in containers. Growing my own shiso at home reminds me of my travels in Japan, where it's commonly grown and used fresh in restaurants.

*In Korea, a* Perilla frutescens *variety with bigger, flatter and broader leaves is referred to as* kkaennip *or Korean perilla. It has a stronger aniseed taste and is commonly used to wrap grilled meat.*

*You can collect shiso seeds from the plant, but it's best to let them dry on the plant and then drop onto the soil, as shiso readily self-seeds. Nature is much more adept at germinating shiso seeds than you or I will ever be!*

# VARIETIES

In Japanese cooking, the frillier-leafed varieties of shiso are used. The **green-leaf variety** is most commonly used for cooking and garnishes, as it has a better flavour. The **red/purple leaf variety** is more bitter. It's used to add colour to pickles (such as umeboshi), and to make perilla juice, a popular Japanese summer drink.

# GROWING

Sow shiso seeds at the beginning of spring. Despite their small size (they look like mustard seeds), they have a hard shell and need to be soaked in water for 24 hours before sowing. They also require light to germinate, so sow on the surface of the growing medium, then press down so the seeds make good contact. Use a heat mat to assist with germination.

Shiso seeds take up to three weeks to germinate. They need to be kept damp – so, gently mist with water. If you're struggling to germinate seeds, then check with your local gardening community online to see if someone can share seedlings with you (shiso self-seeds easily after the first season!). Germinating seeds is the most difficult part of growing shiso – once it's done, things are a lot easier.

Transplant the seedlings when they have three or four adult leaves. Space the plants 30–40 centimetres (12–16 inches) apart; they will also happily live in pots that are at least 20 centimetres (8 inches) in diameter.

Shiso is drought tolerant when established, but prefers to be watered well. Plants will grow all through summer and autumn, and will die down in cooler climates once the weather cools down.

# HARVESTING

Give your plant a couple of months to establish itself. Once it's a reasonable size, then you can start harvesting the leaves as you need them.

Encourage your shiso to bush up by snipping off the growing tips just above the last two leaves. The plant will grow two new stems to replace the one you snipped off, resulting in a bushier plant.

# STORING

Shiso leaves can be stored in an airtight container in the fridge for up to four days.

# COOKING

There are many ways to use shiso in dishes:
- Make pesto, but replace the basil leaves with shiso leaves.
- Garnish poke bowls, sashimi or chirashi bowls.
- Roll half a leaf into the filling of a sushi roll, rice paper roll or spring roll.
- Add it to cold drinks, instead of mint.
- Include it in salads for a minty touch.

# SPRING ONION (SCALLION)

*Allium fistulosum*
Mandarin: 葱 – cōng
bunching onion · green onion · salad onion · Welsh onion

| | | |
|---|---|---|
| | **PLANT TYPE** | Perennial |
| | **PLANT FAMILY** | Onion family (Alliaceae) |
| | **PLANT SIZE** | Upright plant to 50 cm (20 in) tall, 5–10 cm (2–4 in) wide |
| | **POT FRIENDLY?** | Yes, plant 3–5 cm (1–2 in) apart and keep them well-watered |
| | **SUN** | Full sun or part-sun |
| | **FROST** | Frost hardy |
| | **WATER** | Regular watering |
| | **FOOD** | Light feeder |

The humble spring onion is possibly the easiest and quickest thing to grow. You could place spring onion bottoms in a glass of water, and they'd start growing fresh stems the same day. (Store-bought spring onions often come with the roots still attached, so all they have to do is pop out some new stems!)

Spring onions are one of the most popular herbs in Asian cooking and regularly used as a garnish. Unlike most other edibles in the onion family (Alliaceae), spring onions do not form a bulb at the base. Instead, they are grown for their edible hollow stems, which have a light, sweet, oniony flavour.

Supermarkets tend to sell spring onions in a large bunch, but I only ever need a few. So, I grow my own! This way, I can easily pop out into the garden and harvest as many as I need, whenever I need them.

*As a member of the onion family (Alliaceae), spring onions make great companion plants. Their pungent stems help to repel pests and mask the scents of other plants.*

## VARIETIES

If you're cultivating spring onions using supermarket-bought ones (see below), then you'll have to settle for the available variety! Otherwise, select a variety based on what's most suited to your climate and preferences.

## GROWING

You can grow spring onions from seed, but I highly recommend the alternative method of growing from supermarket stock.

Buy a bunch of spring onions with the roots still attached. Cut off the green stems, leaving approximately 5 centimetres (2 inches) of the stem bases still attached to the roots. Plant these roots straight into damp soil, and water in well to allow the plants to settle. They will quickly grow fresh stems.

If you're not ready to plant out your spring onions on the day you buy them, then pop them into a glass of room-temperature water, leaving the tips of the shortened stems above the waterline. Place the glass on a windowsill. The stems will grow before your eyes! Plants can be left to grow in the water for a couple of weeks before they lose vigour and should be planted out. Be sure to change the water every few days, or it will start to get slimy and smell.

The best time to start new plants is spring because, even though they grow year-round, they grow fastest when the weather warms up and slowest when the weather is cool. Plant your spring onions in a sheltered position away from strong winds, as the stems can easily bend. Full sun and part-sun are both okay.

You can plant them 3–5 centimetres (1–2 inches) apart, as they are small plants with small roots, and will happily crowd together. However, planting 5–10 centimetres (2–4 inches) apart will give them more space and nutrients. Eventually, they'll grow to the size of the original spring onions that you cut and planted.

You can easily grow spring onions in a pot. The best way is to grow several together – spaced 3–5 centimetres (1–2 inches) apart – in a large pot, rather than planting them in individual pots. This is because the growing medium in larger pots stays moist for longer, which is important for spring onions. They will also happily grow indoors next to a sunny window.

Spring onions are light feeders, so they don't need as much feeding as other plants. However, because they have small root systems, they need to be kept well-watered and are prone to drying out in warm weather.

## HARVESTING

Supermarket growers harvest spring onions by pulling the entire plant out of the ground. However, for home growers like us, it's best to harvest the stems using the 'cut and come again' method. Simply snip off (at the base) as many individual stems as you need. You can also cut the entire plant down to the base, and it will regrow.

If your plant flowers, then snip off the blooms to redirect the plant's energy towards growing the stems, or simply cut the entire plant at the base and let it regrow. Alternatively, let it flower to feed the bees. The flower buds and little flowers are both edible, and they're very pretty when used as a salad garnish.

## STORING

To keep spring onions crisp and fresh for weeks in the fridge, all you have to do is cut them to fit into an airtight container, with a cloth napkin on the bottom.

## COOKING

Most of the time I use spring onions as a garnish in my cooking. They add a fresh, garden-to-table touch to whatever I'm making, and seem to work with almost anything. I simply cut off the number of stems I need, give them a rinse and then roughly chop them up. Depending on my mood, I'll use either kitchen scissors or a knife – both work.

During winter, my husband, Tom, loves making congee on a weekly basis, garnishing it with large handfuls of spring onion from our garden. We have dozens of plants, but because they grow slowly at that time of year, they barely keep up with his garnishing needs!

For the times when the garden gifts me with a bounty of spring onions, I'll make one of these two dishes: our family's spring onion pancake recipe, or two-minute silken tofu (see page 252).

### Our family's spring onion pancake recipe

A popular Chinese street food, spring onion pancakes can sometimes be found on restaurant menus as an appetiser. I vividly remember helping my mum make large batches of these savoury pancakes as a kid. They taste amazing: salty, crispy and flaky when you rip into them - like well-made croissants. She's lovingly let me share her secret recipe, so you can make them too.

The recipe yields four spring onion pancakes; if you have lots of spring onions (scallions), then you can make a big batch to freeze. Just stack the uncooked pancakes in a cake box with compostable baking paper between them. The pancakes can be cooked from frozen in the same way as fresh (just cook them for longer). Keep in mind that, when making bigger batches, you'll need to roll out the dough into a larger rectangle. So, I suggest doubling the recipe at most and making multiple double batches.

248

**MAKES 4 PANCAKES**

250 g (9 oz) plain (all-purpose) flour (plus extra for dusting)

170 ml (5¾ fl oz) freshly boiled water

80 g (3 oz) spring onion (scalllon) greens

3½ tablespoons vegetable oil

1 teaspoon cooking salt

Place the flour into a large mixing bowl. Slowly pour the freshly boiled water into the bowl, mixing at the same time. Knead together until it forms a smooth dough. Use a little flour if needed to stop the dough from sticking to your hands. Cover the bowl, and set it aside for 1 hour.

While the dough is resting, finely chop the spring onion greens.

Once the dough is ready, lightly dust the work surface with flour. Transfer the dough to the work surface, and knead for 1 minute.

Before rolling, generously dust the work surface with flour, to prevent the dough from sticking. Using a rolling pin, roll the dough into a rectangle 60 × 30 centimetres (24 × 12 inches) in size, and approximately 3 millimetres ($^1/_{10}$ inch) thick.

Combine 1 tablespoon flour, 1½ tablespoons vegetable oil and the cooking salt in a bowl. Break up any lumps of flour with the back of a spoon. Pour the flour, vegetable oil and salt mixture all over the dough – use the back of a spoon or a pastry brush to help spread the oil – and let it soak in for 1 minute. Spread the spring onion greens evenly on top.

Roll up the dough to form a baguette shape, gently tucking in the edges to stop spring onion greens from falling out of the sides. If the dough sticks to the work surface, then use a dough scraper. Cut into 4 even sections.

Pinch shut the ends of each piece of dough so no spring onion greens fall out. While holding on to the ends, gently knock each piece of dough against the work surface while stretching it out into a longer roll.

Twist the dough, then roll it into a scroll like a snail shell. Tuck in the ends. Repeat the process for each piece of dough.

To roll out the pancakes, dust the work surface with more flour. Using a rolling pin, roll out each scroll gently to create pancakes approximately 15–20 centimetres (6–8 inches) in diameter and 5 millimetres ($^1/_5$ inch) thick.

To cook the pancakes, heat 2 teaspoons vegetable oil in a sauté pan over medium heat. Place one spring onion pancake in the pan. Cover with a lid, and cook for 3 minutes or until it starts to brown. Flip the pancake, cover with a lid, and cook for a further 3 minutes or until it starts to brown. Repeat for each pancake.

Serve the pancakes fresh out of the pan.

*When making spring onion pancakes, it's best to use only the green parts of the stem because they're thinner and mix into the dough better. The thicker white stalks can be used to flavour soups or to make veggie stock.*

2

250

## Two-minute silken tofu

This is one of those dishes I make whenever I can't be bothered cooking. It's a soft tofu that's served raw, topped with a few simple ingredients and covered with homegrown spring onions (scallions). The recipe calls for one spring onion, but you can add as many as you like. The dish takes a couple of minutes to prepare and is rich in flavour, despite the minimal ingredients.

Chop the spring onion.

Prepare the tofu by removing the plastic film on top of the tofu box and draining any excess water. While the tofu is still in its box, slice it into 1 centimetre (½ inch) squares. Make sure that you slice all the way to the bottom.

Cover the tofu box with your serving plate, then gently and quickly tip it over so the sliced tofu comes out, still retaining its shape.

Top the tofu with the ingredients in the following order: light soy sauce, zha cai, spring onion and sesame oil.

Serve immediately, as tofu starts to release water if it's left to sit.

**SERVES 2**

1 spring onion (scallion)

300 g (10½ oz) silken tofu block

1 teaspoon light soy sauce

2 tablespoons zha cai (see page 73)

1 teaspoon sesame oil

252

*Silken tofu and zha cai are two staples that we always have in our fridge, as both keep fresh for a few weeks and always come to the rescue on days when the menu is light.*

# THAI BASIL

*Ocimum basilicum*
Mandarin: 泰国罗勒 – tài guó luó lè
anise basil • Asian sweet basil • horapha

| | PLANT TYPE | Perennial; grown as an annual in cool climates |
|---|---|---|
| | PLANT FAMILY | Mint family (Lamiaceae) |
| | PLANT SIZE | 50 cm (20 in) tall, 30 cm (12 in) wide |
| | POT FRIENDLY? | Yes, plant one in a 20–30 cm (8–12 in) pot |
| | SUN | Part-sun |
| | FROST | Frost sensitive |
| | WATER | Regular watering |
| | FOOD | Light feeder |

Thai basil tastes quite different from Mediterranean basil. Rather than being sweet, it has a spicy, aniseed and mint flavour. These aromatic plants have deep purple stems and flowers, which are a striking contrast against the shiny green leaves. Just like all other basils, they have a bushy habit, are easy to grow and can give you a continual harvest all season long.

In warmer climates, Thai basil can be grown as a perennial. This is a bonus, as you don't have to sow it yearly like sweet basil, which is an annual.

# GROWING

The best time to start Thai basil plants in cooler climates is early spring. In warm, frost-free climates, you can start at any time of the year.

I prefer to grow Thai basil from a punnet of seedlings, as I've found a nursery in Melbourne that stocks a wonderful variety of basil. If your local nursery doesn't have it, then it's also possible to grow Thai basil from the herb itself. Buy a bunch of fresh Thai basil from the shop, remove the bottom leaves, and pop the bunch into a jar of clean water. When roots form, transplant outdoors after the last frost.

If you want to raise Thai basil from seed, then sow the seeds on the surface of the growing medium and cover with vermiculite. Be sure to use a heat mat, as Thai basil seeds need a warm temperature to germinate. Expect seedlings to emerge around a week later; transplant them into the garden after the last frost, when the seedlings have at least three leaves.

Plant your Thai basil in a part-sun spot in the garden, as it has delicate leaves that can be burned by the sun on hot days. Thai basil will also happily grow in a pot. The bigger the pot, the bigger the plant will grow – so choose the size according to how much Thai basil you want. You can even grow it indoors near a bright window – I've seen friends do this quite successfully.

Thai basil tends to flower quite early, usually around mid-summer. If you let it flower, then it'll focus on the flowers and forget about its foliage. The leaves end up growing smaller and drier, and the stems become woody. So, to extend your herb harvest, snip off the flower buds as soon as you see them.

Come autumn, when the weather cools down, you can continue to grow the plant as a perennial if you're able to protect it from frost. Otherwise, plant again next spring.

*Thai basil's strong fragrance is said to repel unhelpful bugs (such as aphids, flies and mites), making it a great companion plant in your veggie patch.*

*Since it's sensitive to full sun and high heat, it's the perfect companion plant for shady spots under fruit trees or climbing beans.*

## HARVESTING

There's a special method for harvesting all kinds of basil that will help you grow more basil. Always harvest plants by snipping off the main stem, just above a pair of leaves. Two small side shoots will then grow from the base of those leaves and help to create a bushier plant.

Regularly harvesting from your Thai basil plant will also delay flowering. Any flowers or flower buds that you do snip off can be used in exactly the same way as the leaves, and they also make a beautiful garnish for salads.

## STORING

Thai basil can be easily frozen and used throughout the year. Simply wash and dry the leaves, place them in a thin layer on a baking tray in the freezer, and transfer them to a re-usable freezer bag once frozen.

You can also dry Thai basil like you would any other basil. Remove the leaves from the bottom 5 centimetres (2 inches) of each stem, tie up the stems in small bunches of approximately ten, and hang them upside down in a ventilated area out of direct sunlight. I like to hang mine indoors, off a clothes rack like laundry – it works well! Leave the bunches for one or two weeks, until the leaves are completely dry and crumble between your fingers. Keep the bunches hanging until you need them, or place the dry leaves into an airtight container and store in the pantry.

## COOKING

Thai basil has an awesome superpower that most basils don't have: it holds its flavour well when cooked. Here are some fabulous ideas for using tasty Thai basil in your dishes:

- Add it to **green and red Thai curries**, stir-fries and soups for an aromatic finish.
- Make **Thai pad kee mao** (Thai drunken noodles).
- Whip up a homemade version of **Vietnamese pho** using beef bone broth, onions, flat rice noodles (banh pho) and beef slices. Top with your homegrown Thai basil, blanched mung bean sprouts, a lemon wedge and fresh chillies.
- Make **Taiwanese san bei ji** (three cup chicken).
- Enjoy it on a **Vietnamese banh xeo**, a savoury crepe with pork, prawn, onion and blanched mung bean sprouts.
- Add it to the filling of **spring rolls or rice paper rolls**.

# VIETNAMESE MINT

*Persicaria odorata*
Mandarin: 越南薄荷 – yuè nán bò hé
Cambodian mint • daun kesum • laksa leaf • rau ram • Vietnamese coriander

258

| | | |
|---|---|---|
| | PLANT TYPE | Perennial; grown as an annual in cool climates |
| | PLANT FAMILY | Knotweed family (Polygonaceae) |
| | PLANT SIZE | 15–45 cm (6–18 in) tall, 15–45 cm (6–18 in) wide |
| | POT FRIENDLY? | Yes, plant one in a 20–30 cm (8–12 in) pot |
| | SUN | Full sun or part-sun |
| | FROST | Frost sensitive |
| | WATER | Keep consistently moist |
| | FOOD | Light feeder |

Vietnamese mint has a tangy flavour like coriander (cilantro), but with a spicy kick. The leaves are commonly used in Vietnamese cooking, as well as dishes from nearby countries (they're essential for making Malaysian laksa, which is why the plant is also known as laksa leaf). The long, ornamental leaves have distinctive red markings that are more pronounced when grown in full sun. While the plant isn't related to common mint (and is not even in the same family), it does resemble mint and grows in a similar way.

Vietnamese mint grows wild in subtropical and tropical Southeast Asia, but it can also be grown in cooler climates as an annual protected from frost. In warm climates, it makes a good substitute for coriander (cilantro) and is a perennial that you only have to plant once.

*A simple way to create additional Vietnamese mint plants is to bury part of a stem under soil and wait for the buried part to root. Once roots form, cut off the segment and pot it up.*

# GROWING

The easiest ways to start growing Vietnamese mint are to:
- buy a baby plant from a garden centre or nursery
- cultivate a stem cutting taken from a friend's plant
- grow from Vietnamese mint bought from an Asian grocer.

Vietnamese mint roots extremely easily. All you need to do is take a 20–30 centimetre (8–12 inch) piece, remove two-thirds of the leaves, and pop the stem into a jar of water. Be sure to change the water every few days to keep it fresh. Transplant into a pot once roots have grown.

Vietnamese mint can spread like regular mint – although it's easier to pull out. It creeps along the ground and will root where its stems touch the soil. Grow it in a pot to prevent it from taking up more space than you want it to. Smaller pots are fine, but bigger pots will give you more leaves to harvest.

In cooler climates, a position in full sun is best. In warmer climates, part-sun or morning sun offers more than enough light and warmth. Vietnamese mint loves water, so keep the soil consistently moist. Mine happily lives in a self-watering pot. Keep it watered well during summer, and protect it from frost when the weather cools. Other than that, there's not much else to do but to enjoy the harvest!

# HARVESTING

Once plants grow to a reasonable size, start harvesting by cutting stems near the base. This encourages the plant to grow back bushier. It'll still regrow even if you cut the whole plant to the ground.

**Pruning** It's best to prune your plant now and then to keep the growth young and fresh. Around winter, cut it back to the ground. This encourages new growth in spring and keeps the plant compact and bushy.

# STORING

Store Vietnamese mint unwashed in a re-usable container in the fridge for up to a week. Alternatively, store it upright in a jar of water in the fridge.

# COOKING

You can use Vietnamese mint in the same way as coriander (cilantro). Add it to laksas, curries and fillings for rice paper rolls. It pairs beautifully with seafood and chicken, and works well in stir-fries and salads.

# FRUITS

# CUMQUAT

*Citrus japonica* (syn. *Fortunella japonica*)
Mandarin: 金柑 – jīn jú
kumquat

| | | |
|---|---|---|
| | **PLANT TYPE** | Evergreen perennial |
| | **PLANT FAMILY** | Rue family (Rutaceae) |
| | **PLANT SIZE** | Up to 5 m (16 ft) tall and 1.5 m (5 ft) wide but can be pruned to 2 × 2 m (7 × 7 ft) |
| | **POT FRIENDLY?** | Yes, plant one tree in a 40–50 cm (16–20 in) pot |
| | **SUN** | Full sun |
| | **FROST** | Frost tolerant |
| | **WATER** | Regular watering |
| | **FOOD** | Heavy feeder |

My parents grow a cumquat tree by their front door. It was a gift from me and my sister for my dad's birthday one year, and he's lovingly cared for it ever since. In many Asian cultures, the cumquat tree is a symbol of good luck.

If you want to dip your toe into growing Asian fruits, then a cumquat tree is a great place to start. It's one of the easiest citrus trees to grow, being more frost hardy than others, and it's a small tree that will happily live in a pot for its entire life without much pruning. Its glossy green leaves make it a popular ornamental tree - many people grow a cumquat simply for its looks. In summer, it boasts small, white, fragrant flowers that give off a sweet citrus scent every time you walk past.

Cumquats look like mini mandarins - but you eat them whole. The edible skin is the sweet part of the cumquat, which pairs with the tangy, tart flesh inside to give you a perfect balance of flavours. It tastes like a vitamin C tablet in fruit form and is, in fact, very rich in vitamin C. I like the idea of growing a cumquat tree near our front door, so I can harvest a dose of vitamin C every time I walk past!

## VARIETIES

My dad and I both grow **Meiwa**, which has sweet, round fruits that are usually seedless. I like Meiwa because it's a compact variety that only grows to 2 metres (7 feet) tall in a pot. I gifted my friend **Nagami**, which is the oval cumquat variety – I thought we could each have different types to share! My sister has a mature Nagami in Perth, and we can both attest to its excellent flavour. All cumquats are self-pollinating, so you only need one tree to grow fruits.

## GROWING

I recommend growing your cumquat in a pot. They grow more slowly than other fruit trees and can cope with container living, so you might as well save your in-ground space for something else. A family friend has two happy and healthy cumquat trees around 2 metres (7 feet) tall that have lived in 50 centimetre (20 inch) pots for more than 30 years!

Cumquat trees prefer slightly acidic soil (pH 5.5–6.5). If you're planting in a pot, then look for a potting mix specially formulated for citrus trees.

Cumquats (and all citrus trees) are prone to root rot and require well-draining soil. If you're planning on planting into the ground, then first check your soil drainage by digging a hole and pouring a bucket of water into it. If the water pools and doesn't disappear within 30 seconds, then it's best to plant the tree in a raised mound to assist with drainage. To ensure good drainage in pots, use a regular pot rather than a self-watering pot.

Cumquats are tolerant of cold weather down to −7 degrees Celsius (19 degrees Fahrenheit). If the weather is any colder, then it's best to bring your tree indoors during winter. During the growing season from spring to autumn, cumquats (and all citrus) like to be regularly watered – but let the soil dry out between watering. Growth slows down in winter, so you don't need to water unless there's no rain.

All citrus trees, including cumquats, are hungry plants that need regular feeding from spring to autumn. Feed every four to six weeks during the growing season with compost or an organic-based fertiliser made for citrus or fruit trees.

*Cumquats are popular gifts for loved ones during Lunar New Year, in order to bring more abundance and prosperity into the recipients' lives.*

## HARVESTING

Cumquats flower later than most other citrus, usually from late spring or summer. These flowers turn into fruits, which are usually ready to harvest from late winter to early summer. You'll know that your cumquats are ready to be harvested when the skin is fully coloured and the fruit feels firm to the touch.

Harvest cumquats with a bit of stem attached. This helps to keep them fresh and minimises damage to the tree. Harvest only ripe cumquats, as they won't ripen further after they're picked.

**Pruning** Cumquats don't need much pruning, as they are already compact, slow-growing plants. They also seem to have a knack for growing in a beautifully well-balanced shape without much intervention.

If you want to prune your cumquat tree, then do it right after all of the fruits have been harvested. Luckily, cumquat (and all citrus) trees require no fancy pruning techniques, unlike stone fruits. Just grab a pair of secateurs, and follow my easy guide:

1. Remove any dead, diseased, damaged or weak branches.
2. Remove any branches that are pointing inwards (as these branches get shaded by other branches and won't develop good fruits).
3. Remove any crisscrossing branches and those that rub against each other (this is to improve ventilation within the tree and to reduce bark damage that may allow diseases to enter the tree).
4. Remove any branches that are growing too low on the trunk.
5. Prune to shape by reducing the length of any branches that are too long by up to a half – if there are new flower buds already forming, then prune around them or you'll lose next season's fruits!

## STORING

Cumquats can be stored in the fridge for one to two weeks. Pop them into your crisper like you would any other citrus fruits. If you want to store them for longer, then you can freeze them whole in re-usable freezer bags. They'll happily keep in the freezer for up to six months.

## COOKING

Most of the time, I eat my cumquats fresh off the tree. A popular way to use a harvest of cumquats is to make marmalade, but if you've already made enough marmalade to last you for years, then why not try one of the following recipe ideas.

## Cumquat syrup

This syrup is traditionally used to fight colds and coughs. However, its delicious taste shouldn't just be reserved for sick days. Mix 1–2 tablespoons of cumquat syrup with 1 cup (250 millilitres) of hot water for a delightful drink, or drizzle it on pancakes, muesli (granola), porridge or ice-cream. To store the syrup, you'll need a clean and sterilised glass jar with a lid (see page 72).

Thinly slice the ginger. Wash the cumquats, remove the little green stems, and cut the cumquats in half. Remove any seeds by squeezing each half over a small strainer sitting above a bowl. The strainer will catch the seeds, and the juice will flow through.

Place the cumquat halves and juice, water, sugar and ginger into a small saucepan. Bring to a boil, stirring until the sugar has dissolved.

Reduce the heat to medium, and simmer for 20–25 minutes or until the cumquats are tender and translucent, and the liquid has thickened.

Allow the mixture to cool down a little before transferring everything (cumquats and syrup!) to the jar. Store it in the fridge for up to 2 weeks.

**MAKES APPROX. 600 ML (20 FL OZ)**

1 cm (½ in) piece of ginger

.............................................

500 g (17½ oz) cumquats (approx. 50–60)

.............................................

1 cup (250 ml) water

.............................................

1 cup (210 g) sugar

.............................................

## Cumquat and lavender iced tea

This refreshing iced tea is inspired by a delicious one that I had while travelling. Mint and lavender are in season at the same time as cumquat, so they're great additions to this iced tea, but you can use any other seasonal garnishes that you might have in your garden.

Slice the cumquats in half and remove the seeds. Tear the mint leaves into small pieces.

Steep the loose-leaf green tea (or tea bags) in 1 cup (250 millilitres) water at 80 degrees Celsius (180 degrees Fahrenheit) for 4 minutes, then remove the tea leaves. Add 2 cups (500 millilitres) cold water to dilute and cool the tea, and add the slice of lemon.

Prepare 4 glasses. Fill the bottom quarter of each glass with ice cubes, then add 2 tablespoons cumquat syrup and 2 cumquat halves. Divide the green tea between the glasses.

Add mint and lavender to each glass, and serve cold.

**SERVES 4**

4 fresh cumquats

.............................................

10 mint leaves

.............................................

1 tablespoon loose-leaf green tea (or 3–4 tea bags)

.............................................

3 cups (750 ml) water

.............................................

1 slice of lemon

.............................................

Ice cubes, for serving

.............................................

8 tablespoons cumquat syrup (see recipe above)

.............................................

4 lavender buds

.............................................

# GOJI BERRY

*Lycium barbarum*
Mandarin: 枸杞 – gŏu qǐ
matrimony vine • wolfberry

| | | |
|---|---|---|
| | **PLANT TYPE** | Deciduous perennial |
| | **PLANT FAMILY** | Nightshade family (Solanaceae) |
| | **PLANT SIZE** | Vining shrub that can grow up to 4 m (13 ft) but can be pruned back to 2 × 2 m (7 × 7 ft) |
| | **POT FRIENDLY?** | Yes, and recommended; plant one shrub in a 30–50 cm (12–20 in) pot |
| | **SUN** | Full sun or part-sun |
| | **FROST** | Frosty hardy |
| | **WATER** | Regular watering but drought tolerant once established |
| | **FOOD** | Light feeder |

Native to China, goji berries are small with glossy red skin, and they have a delicate, sweet, herbal taste similar to a cranberry. They are rich in vitamins and antioxidants. As it's hard to find fresh goji berries at the supermarket, you'll have to grow your own.

The plant is very low maintenance and is both frost hardy and drought tolerant once established (it prefers hot, dry summers). It grows into an interesting shape, with long vines that form a weeping habit. The vibrant red berries that hang off the plant look a bit like those of Christmas holly.

The best thing about growing goji berries is the long harvest season. Berries start ripening in summer, and the plant will fruit continually until it is hit by frost. This is perfect for home gardeners, as you have more time to enjoy the harvest.

## VARIETIES

Both **Lycium barbarum** and **L. chinense** are known as goji berries. However, L. barbarum is the species you should grow if you want to eat the berries. The berries of L. barbarum are larger than those of L. chinense (which is grown more for its leaves) and twice as sweet.

## GROWING

Aside from buying a goji berry plant from your local nursery, the easiest way to start a goji berry plant is to take a cutting from an existing plant during winter, while it is dormant. Snip off a 30–40 centimetre (12–16 inch) piece of woody stem, right below a node, making sure that it has at least four nodes. Bury the stem in soil (with at least two nodes under the surface) and keep it moist. The cutting will take quite easily when propagated at this time; you'll know that it's taken when you see new growth emerge in spring.

You can also propagate by layering. Bend a stem of an existing plant downwards and bury it. The part of the stem that touches the soil will grow roots. Once these appear, simply cut off that stem section and pot it up to get a new plant.

Goji berry plants created through cuttings or by layering will start producing berries after two years and reach full capacity at around three to five years. This is much quicker than plants grown from seed – they can take up to four years to produce their first berries. Goji berries are self-fertile, so you only need one plant to get fruits.

Being deciduous plants, goji berries sit dormant throughout winter and then pop out clusters of light purple flowers in late spring. These flowers grow into berries quickly, within five weeks. Berries start green and turn glossy red as they mature. The berries grow on long, flexible stems that start upright but flop over on top, so the stems need to be propped up with stakes or a trellis. Let the stem tops droop like the branches of a weeping willow tree.

### TAKE CARE!

- Goji berry plants spread via underground runners and can pop up around the garden. To stop your plant from becoming weedy (at best) and invasive (at worst), grow it in a large pot.
- Be mindful when transplanting goji berry plants, as they have a long, delicate taproot and thorny vines.

Thankfully, the plants aren't particularly fussy. They tolerate most soils except heavy clay, which doesn't drain well and can cause root rot. Water young plants regularly, until they become established. Mature plants are drought tolerant and much more resilient.

A full-sun position will lead to the best yields, but protect your plants from afternoon sun in warmer climates. The plants will also happily live in part-sun but will fruit less. Feed your goji berry plants once a year at the start of spring.

**Pruning** It's hard to go wrong when pruning goji berries. They are happy to be pruned hard, and pruning can help to increase yields by promoting more new growth in spring.

The best time to prune goji berries is when they're dormant in winter. Remove any dead or damaged growth – any stems that snap off easily are dead – and any crisscrossing branches. Cut the plant down to a manageable size, and thin out branches if there are too many. Winter is also a great time to update your trellis, as it's much easier to see the plant's framework while there are no leaves in the way!

## HARVESTING

Goji berries are ready to harvest when they are bright red and slightly firm when squeezed. Simply pick them off one by one. It can feel a bit tedious if you have a lot to harvest, but the more you pick, the more you get.

## STORING

Goji berries ripen slowly throughout the season, so the harvests are regular and manageable compared to those of berries that ripen on the bush all at once. Freshly picked goji berries can be stored in an airtight container in the fridge for up to a week. The best long-term storage option is to dehydrate them (see box).

272

### DEHYDRATING GOJI BERRIES

Pop them in a single layer on an oven tray, and dry them indoors or outdoors in direct sunlight. Gently shake the tray every now and then so they dry evenly. Alternatively, use an oven on the lowest temperature setting, or a dehydrator at around 50 degrees Celsius (120 degrees Fahrenheit) for eight to ten hours. The berries are dried when they have shrunk and are wrinkly and hard. Place them into an airtight container, and store them in the pantry until next season rolls around. Dried goji berries can be used as they are (or rehydrated first) in any way you would use fresh goji berries, as well as to make homemade trail mixes and muesli (granola).

# COOKING

The subtle sweetness of goji berries works well in both sweet and savoury dishes. In my family, goji berries are usually used in cooked savoury dishes, such as soups and stir-fries. My mum also often has a glass of goji berry tea on the kitchen bench. Goji berries are perfect for desserts, and I think that they're a great ingredient for breakfast dishes – they make a not-too-sweet substitute for sultanas and raisins.

## Summer berry garden infusion

Decorate your drinks with the flavours of summer. Whenever I have a harvest of berries, I love to infuse them like this, capturing the ephemeral and every-changing berry season.

Pop everything into a large jug or water bottle, and enjoy refreshing, berry-infused $H_2O$.

**MAKES 3–4 CUPS (750–1000 ML)**

2 tablespoons fresh goji berries

.........................................................

1 big handful of other seasonal berries from the garden (blueberries, raspberries, blackberries)

.........................................................

2 cups (500 ml) water

.........................................................

3–4 ice cubes

.........................................................

3–4 mint leaves

.........................................................

### MORE GOJI GOODNESS

Here are some other ideas for using freshly picked goji berries:

- Add them to bone broth, veggie broth or soups (such as shark fin melon soup; see page 172) for a hint of sweetness. A good amount is 1 tablespoon of goji berries in a four-serving soup.
- Use them to replace sugar in savoury stir-fries.
- Mix them with leafy greens for a vibrant pop of colour.
- Use them in place of raisins when baking things such as muffins or cookies.
- Brew tea by infusing 1 tablespoon of goji berries in 1 cup (250 millilitres) of hot water for five to seven minutes. You can also add ginger, jujubes or chrysanthemum flowers, as well as green or oolong tea.
- Throw some into a glass of lemonade.
- Sprinkle them over fresh salads to add an extra layer of flavour.
- Add them to summer smoothies.
- Enjoy them fresh as a healthy snack.
- Use them as a porridge or oatmeal topping.

# JUJUBE

*Ziziphus jujuba*
Mandarin: 红枣 – hóng zǎo
Chinese date • Chinese red date • red date

| | | |
|---|---|---|
| | **PLANT TYPE** | Deciduous perennial |
| | **PLANT FAMILY** | Buckthorn family (Rhamnaceae) |
| | **PLANT SIZE** | Up to 10 m (33 ft) tall and 5 m (16 ft) wide but can be pruned to 2 × 2 m (7 × 7 ft) |
| | **POT FRIENDLY?** | Yes, and recommended; plant one tree in a 40–50 cm (16–20 in) pot |
| | **SUN** | Full sun |
| | **FROST** | Frost hardy |
| | **WATER** | Light to medium watering |
| | **FOOD** | Light to medium feeder |

One of my best edible garden discoveries has been jujube, a tree that seems like it grows two different fruits in one. Jujube fruits look like mini apples and can be harvested at any stage of maturity, with each stage giving you a different flavour profile.

When fresh and green, jujubes are crisp and taste similar to apples. When dried, they're deep red in colour with a rich, sweet taste and chewy texture.

Native to China, jujube is one of the oldest cultivated fruit trees. It is an extremely hardy plant and tolerates frost, high heat and drought. Jujube trees have a narrow, upright growth habit with weeping leaves, and they make a great edible substitute for an ornamental weeping cherry. Just be sure to put your jujube in a pot, as it grows enthusiastically and can send out suckers and become invasive.

## VARIETIES

Choose your jujube based on whether you prefer it fresh or dried. Some varieties are better enjoyed fresh (**Li**, **Silverhill** and **Sihong**), while other varieties are better dried (**Lang**, **Chico** and **Ta-Jan**). Varieties that are better for drying have a lower water content, so aren't as juicy when fresh. But all varieties can be eaten either way.

There's a saying that 'if you can only fit one jujube, then make it a Li'; it's the variety I grow. It has large, sweet fruits with small pits.

Most jujubes are considered self-fertile, so you need only one tree to get fruits. However, you'll get even more fruits off your tree if there's a second tree nearby.

## GROWING

Grafted jujube trees are usually available at nurseries in winter while they're dormant. This is also the best time to transplant or repot your jujube tree.

A jujube tree's dream home is a temperate climate, somewhere with long, hot, dry summers and cool winters with 150–400 chill hours (this is where the temperature is under 7 degrees Celsius/ 45 degrees Fahrenheit). Humidity can increase the risk of disease and make it harder to dry jujubes on the tree.

Jujubes prefer a full-sun position, away from strong winds that can damage the plant. They're pretty tough trees and will tolerate a wide range of soils. Jujubes grow a deep taproot and are drought tolerant once established, but they're best watered (especially when fruiting) if you want a good yield. They're also quite low maintenance food-wise, but feeding once a year when they start to flower will help them grow more fruits.

It's best to grow your jujube in a pot. When grown in the ground, jujube trees can send out root suckers up to 10 metres (33 feet) away, which you and your neighbours may not appreciate. If you do have a big space and want to grow a jujube tree in the ground, then it can be pruned to 2–3 metres (7–10 feet) in height to make it easier to reach the fruits.

## HARVESTING

Grafted jujube trees love to fruit in their first year. It's best to remove these fruits when they're immature to encourage the tree to establish solid roots and a branching framework. (Maybe allow one or two fruits to mature on the tree to give you a preview of what's to come in future years!)

Jujube trees will flower over a period of two months in late spring to summer. These flowers then turn into fruits that are ready to pick in late summer to autumn. Each light green flower is rather insignificant – it's less than 1 centimetre (½ inch) in size and blooms for only one to two days. However, the flowers bloom successively, which means that the fruits ripen in the same way, giving you a long harvest season.

Jujubes harvested **green and firm** are eaten fresh and taste like a mild, less juicy apple. They make a great summer substitute for apples in salads. I personally love snacking on them at this stage while pottering around in the garden. Jujubes picked at this stage won't continue to ripen after harvest.

Jujubes harvested **half green and half red** or **red but still firm** are also eaten fresh and taste sweeter than green and firm jujubes. They will continue to ripen after harvest.

If you're growing jujubes to use dried, then let the fruits hang on the tree until they're fully red and starting to wrinkle. At this stage, they are at their sweetest. From there, you can either let them sun-dry on the tree if your climate allows it, or bring them inside and dry them in a dehydrator – just don't forget to remove their pits.

**Pruning** Jujubes have unique pruning needs. Unlike most fruit trees that respond to a single pruning cut by sending out branches just below the place where the cut was made, jujubes follow a special rule known as 'one cut stops, two cuts sprout'. This means that you'll need to make two cuts to your jujube to encourage it to branch: one to the trunk, and one to the branch below it.

A jujube tree has one primary (main) trunk that grows vertically, and it sends out secondary (lateral) branches as it grows. These secondary branches bear the fruiting spurs (little nodes/bumps across the branch) from which the deciduous shoots (branchlets) grow and form fruits. When you prune to maintain your tree's size, remember to shorten only secondary branches, and leave the little knobbly fruiting spurs on the pruned tree alone – as unsightly as they look!

Jujube trees are typically grafted onto a rootstock that easily suckers – it's common to find little shoots growing from below the grafting union (at the base of the tree). Pluck them off as soon as you see them, as they can dominate and take over the tree.

## STORING

Fresh jujubes can be stored in the fridge for two weeks. Dried jujubes can be stored in an airtight container in the pantry for up to six months.

## COOKING

Mature, wrinkly jujubes are great for cooking and are often used in a similar way to goji berries to add a hint of sweetness. Use them in:

- my shark fin melon soup recipe (see page 172)
- bone broth soups
- your morning porridge
- sweet Asian dessert soups
- tea with ginger and goji berries
- stir-fried dishes.

**PRUNING TECHNIQUE**

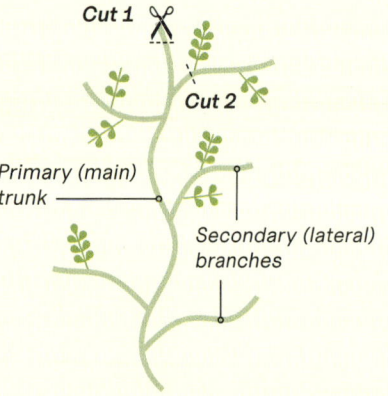

Cut 1

Cut 2

Primary (main) trunk

Secondary (lateral) branches

*When harvesting jujubes to eat fresh, it's best to leave a bit of stem attached to the fruits to help them store for longer.*

# LOQUAT

*Eriobotrya japonica*
Mandarin: 枇杷 – pí pá
biwa • Japanese medlar • Japanese plum

| | | |
|---|---|---|
| PLANT TYPE | Evergreen perennial | |
| PLANT FAMILY | Rose family (Rosaceae) | |
| PLANT SIZE | Up to 7 m (23 ft) tall and 5 m (16 ft) wide but can be pruned to 2 × 2 m (7 × 7 ft) | |
| POT FRIENDLY? | Only while it's young; best grown in the ground | |
| SUN | Full sun or part-sun | |
| FROST | Frost tolerant | |
| WATER | Regular watering; drought tolerant once established | |
| FOOD | Medium feeder | |

When I was a kid, I would follow a special route on the way home from school that took me past all of the local loquat trees. I would stuff my pockets with my foraged bounty and loved devouring these sweet fruits as I walked. In Melbourne, loquat trees hang over back fences, thrive in front yards and even grow in public spaces. Many gardeners plant seeds from the loquats they eat, and sometimes the seeds are dropped by birds and then happily grow on their own. It's a testament to how easy it is to grow a loquat tree.

Loquats are native to the cool, hilly areas of China and have been grown in Japan for more than a thousand years. The trees remind me of frangipanis but with edible fruits. They have large, glossy leaves and sweet, fragrant, white flowers that bloom from late autumn to winter.

I love loquat trees because they fill a gap between other fruiting trees and offer a welcome early summer harvest after months of citrus, apples and pears. The fruit looks like a small, yellow apricot and tastes like a tropical, pineapple-y stone fruit. Loquat fruits aren't typically sold commercially, as they bruise easily so don't travel well. But in a home garden, a loquat tree will soon add a special tropical sparkle to your harvest.

## VARIETIES

**Nagasakiwase** is a popular early-season variety with very sweet fruits. Other interesting varieties include **Bessel Brown**, **Enormity**, **Champagne**, **Sewells Prolific** and **Herds Mammoth**. Loquats are self-fertile, so you need only one tree to get fruits. Keep an eye out for trees on dwarf rootstock, as they are best suited to smaller gardens.

## GROWING

Loquat trees can be grown from seed, but this is a lottery. These trees take up to ten years to fruit, they don't always produce the same fruits as the parent plant, and the fruits sometimes taste bitter or have more seeds than flesh. If you've ever been foraging for loquats and noticed that some trees have fruits that aren't as tasty as others, then this is the reason. I think that if you're going to plant one backyard loquat, then you might as well do it properly. Pick up a grafted, named variety from your local nursery (this is different from buying a loquat seedling from the nursery, which would yield the same results as planting your own seeds, as mentioned above). Grafted loquat trees produce fruits within two years, and often have sweeter fruits with a better flesh to seed ratio.

Loquat is native to temperate and subtropical areas. Although the mature trees are frost tolerant, the baby trees, flowers and fruits are not. Hence, they're ideally grown in climates with a low chance of frost, or in a spot that can be protected from frost.

While I'm a big fan of fruit trees in pots, unfortunately I don't recommend growing loquat in a pot (at least, not as its forever home). This is because loquats grow quite quickly and can soon become root-bound (this means the roots have grown so much that they've filled up the entire pot), reducing their productivity. On the bright side, loquats grow well in the ground and are happy to be pruned back heavily to a compact size.

When choosing a location for your loquat tree, keep in mind that it is an evergreen tree with large leaves that will shade other plants in your garden. I recommend locating it at the back of your garden, where it will be in full sun but won't overshadow other plants.

*Loquat's large, glossy leaves make it a beautiful tree choice for a front yard. Consider it a more cold-tolerant and edible substitute for a frangipani.*

Loquats will adapt to most soil types. Young plants require regular water, but older plants are fairly drought tolerant once established – this is probably why they survive so well in the wild. Of course, they grow much better if well-watered, especially in summer. Loquats aren't heavy feeders but do appreciate fresh organic matter from time to time.

**Flowers and fruits** Loquats flower in autumn, in clusters on the tips of their branches. If an early frost strikes, the flowers will drop off the tree and you won't get fruits that year. This is totally okay – it's just nature. Enjoy the tree as an ornamental plant for that year – in fact, loquats are often grown as ornamentals as well as fruiting trees. If your tree does retain its flowers, then the fruits will appear around three months later.

## HARVESTING

The fruits are ready to harvest when they've reached their full size – approximately 4 centimetres (1½ inches) in diameter, depending on the variety – and are completely yellow. You can pick them at various stages from there – either firm if you prefer them tart, or soft if you prefer them sweet. I definitely prefer them soft and sweet. If it's your first time eating loquats, then pick a few at various stages to see what you like best. Once picked, they won't ripen further on the bench.

Harvest loquats with a bit of the stalk attached to help them store for longer. Don't pull them off the tree, or you may rip off the stalk and expose the flesh on top of the fruit. If the entire cluster is ripe, I find it easier just to snip it off like a bunch of grapes. Handle the fruits gently, as they're quite delicate and will easily bruise.

**Pruning** I like to keep all of my fruit trees, including loquats, to around 2 metres (7 feet) tall. This means that I can grow more fruit trees in my urban garden, and that the fruits are easy to reach and harvest.

Start by pruning your baby loquat trees to your preferred shape. Once the tree is established, prune yearly in early summer, after all of the fruits have been harvested for the season. Cut back all branches that have fruited, and the tree will then grow new branches for the next season's fruits.

If your tree has too many baby fruits on it, it's best to thin them to four or five fruits per cluster to improve fruit size. Trees that are overburdened with fruits tend to start fruiting biennially.

## STORING

Loquat fruits are best eaten fresh as a seasonal treat, as they bruise easily and don't keep well in the fridge or fruit bowl. If you want to preserve loquats, then you can dehydrate them. This can be done by slicing the fruits in half and removing the seeds. Dry at 50 degrees Celsius (120 degrees Fahrenheit) in your dehydrator (or at the lowest setting in your oven) for six to eight hours or until dry. Once dry, transfer the fruits to an airtight container, and store them in your pantry for up to six months.

## COOKING

Loquat fruits have edible skin and one to four slippery brown seeds in the centre. I personally prefer to peel my loquats before eating them, as the skin is a bit tangy.

If you have a large batch of loquats, you can remove the seeds and slice up the fruits, then:

- use them in baked goods – as a fruit-pie filling, or as an apple replacement to make loquat crumble
- make jam or marmalade
- create fruit leathers
- add them to smoothies
- throw them into a red curry with potatoes and seasonal veggies
- include them in an early summer salsa.

# NASHI PEAR

*Pyrus pyrifolia*
Mandarin: 酥梨 – sū lí
apple pear • Asian pear • Chinese pear • Japanese pear • Korean pear

| | | |
|---|---|---|
| 🌱 | **PLANT TYPE** | Deciduous perennial |
| 🌰 | **PLANT FAMILY** | Rose family (Rosaceae) |
| ⤧ | **PLANT SIZE** | Up to 5 m (16 ft) tall and 4 m (13 ft) wide but can be pruned to 2 × 2 m (7 × 7 ft) |
| 🪴 | **POT FRIENDLY?** | Yes, plant one tree in a 50 cm (20 in) pot, although best grown in the ground |
| ☀ | **SUN** | Full sun |
| ❄ | **FROST** | Frost hardy |
| 💧 | **WATER** | Regular watering |
| ⁖ | **FOOD** | Medium feeder |

Nashi pears are native to eastern Asia and happen to be the most eaten pears in the world. The word *nashi* is Japanese for pear.

Nashi pears look and taste a little different from European pears. They're (usually) round like an apple, have a crisp texture and are sweet and very juicy. (The nashi pears you find in the supermarket tend to be less sweet because they are harvested early.) Traditionally, they were carried around as a portable water bottle in fruit form. To me, eating a nashi pear is the perfect thirst quencher!

These are great beginner Asian fruit trees. They are largely free of diseases and pests, start fruiting at a younger age than European pears, fruit earlier in the season and are more compact in size. They also have a lower chill-hour requirement than European pears, so they're a great alternative for areas that are just a tad too warm for European pears. In spring, they provide an exquisite display of beautiful white flowers that rivals that of a cherry blossom tree. If you love cherry blossoms but wish you could grow something productive, then try a nashi tree – it's a great edible alternative.

## VARIETIES

Some nashi pears are self-fertile, while others require cross-pollination for fruit set. Nashi pear trees can cross-pollinate with other nashis as well as early-season European pears. Keep an eye out for double-grafted trees – I often see nurseries sell two nashi pears on the one tree, or a nashi pear and a European pear on the one tree, where both varieties can pollinate each other.

If you live in a temperate to warm (but not tropical) climate, look for a variety with low chill hours (such as **Ya Li** or **Shinseiki**). Ya Li is extremely sweet and uniquely shaped like a European pear. Other varieties to check out include **Hosui** and **Nijisseiki**. All of these varieties will fruit on their own, but they'll do better with another nashi pear (or early-season European pear) tree nearby. Keep an eye out for trees on dwarf rootstock, as they are best suited to smaller gardens.

## GROWING

Position your nashi pear tree in full sun. It prefers well-drained, slightly acidic soil (pH 6.5) but will tolerate other pH levels as well. Water the tree regularly to help prevent the fruits from cracking.

Nashi pear trees bloom at the start of spring, with large, fluffy clusters that contain up to a dozen blossoms. While the trees themselves are frost hardy, their blossoms can be damaged by late frosts, so in cooler climates choose a location in your garden that offers a bit of frost protection.

A nashi pear tree's overenthusiasm for flowering can become problematic, as each one of those flowers will try to become a fruit. Unfortunately, this is a little optimistic because allowing every flower to continue growing can result in a basketful of undergrown fruits. Also, the extra energy needed to support so many fruits may cause your nashi pear tree to start bearing biennially.

Therefore, it's best to do a little fruit thinning to help the tree along. Once the fruits reach the size of a grape, thin every cluster down to one or two fruits. The remaining fruits will grow bigger, leading to a better harvest overall.

### CHILL HOURS

Nashi pear prefers a temperate climate, as it needs 300–500 chill hours at lower than 7 degrees Celsius (45 degrees Fahrenheit) to set fruit. Unfortunately, this means that if you live in a very warm or tropical climate, then you'll have to give this one a miss!

Few pests and diseases bother nashi pear trees, but keep an eye out for fire blight. This bacterial disease affects new growth, causing leaves to look burned and die, and it can quickly kill the tree. As soon as you see fire blight, prune affected stems at least 30 centimetres (12 inches) below the infection, place all infected material in a sealed garbage bag so the disease doesn't spread, and throw the garbage bag into the rubbish bin (don't compost the infected material). Make sure that you sanitise all tools to prevent spreading the disease.

Nashi pear trees grow quickly and are happy to be pruned hard. Remove dead, diseased and damaged branches, and cut back yearly in winter or summer. Winter pruning is best in the first few years, as it encourages growth. Summer pruning is best for more mature trees, as it results in slower growth.

## HARVESTING

Four to seven months after flowering (depending on the variety), your nashi pears will be ready to pick. Harvest season starts around late summer to late autumn. Fruits taste best when they are left on the tree for as long as possible, as they develop their sweetness during this time. They don't sweeten after harvesting, which is why supermarket nashi pears (picked too early) aren't very sweet. Pick fruits once they've changed to their final colour (yellow, green or brown, depending on the variety you're growing) and start to feel heavy.

## STORING

Nashi pears are fabulous storers and can keep fresh in the fridge for up to two months. They do, however, bruise easily, so treat them gently, like eggs.

## COOKING

I love to eat my nashi pears fresh. They can also be added to salads in place of apples and European pears (try them in a Waldorf salad!) or enjoyed as part of a cheese platter. Here are some further ways to use nashis:

- Add them to my kimchi recipe (see page 121).
- Make baesuk (a Korean dessert featuring poached nashi pears).
- Bake nashi tarts, muffins or crumbles.
- Create a tasty chutney, sweet jam or preserve.
- Transform nashi slices into fabulous fritters.

*At the end of the harvesting season, dehydrate extra nashi pears by slicing them thinly and drying them in a dehydrator or oven at the lowest setting. Once dry, store the slices in an airtight container in the pantry.*

*You can add dried nashi pear to winter porridges or enjoy it as a healthy afternoon snack. In addition, you can dice the dehydrated fruits and use them in your own fruit tea blends.*

# PERSIMMON

*Diospyros kaki*
Mandarin: 柿子 – shì zi
Chinese persimmon • Japanese persimmon • kaki • oriental persimmon

| | PLANT TYPE | Deciduous perennial |
|---|---|---|
| | PLANT FAMILY | Ebony family (Ebenaceae) |
| | PLANT SIZE | Up to 5 m (16 ft) tall and 4 m (13 ft) wide but can be pruned to 2 × 2 m (7 × 7 ft) |
| | POT FRIENDLY? | Yes, plant one tree in a 50 cm (20 in) pot, although best grown in the ground |
| | SUN | Full sun or part-sun |
| | FROST | Frost hardy |
| | WATER | Regular watering |
| | FOOD | Medium feeder |

The persimmon is one of the world's most underrated fruits. It's sweet like honey, and when it ripens in autumn, the flavour gives warmth to the cooling days. The tree is native to China, but it's also been grown in Korea and Japan for thousands of years.

The persimmon tree is one of the most beautiful deciduous fruit trees you can grow, and it looks stunning in a front yard. I think of it as an edible substitute for a maple tree, with large leaves that turn from fiery red to orange and yellow in autumn. The fruits ripen after the leaves have dropped, and they look like little orange lanterns hanging on the tree. My persimmon tree is planted right next to the driveway, so my family (and our neighbours) can enjoy its elegance every day.

*Persimmons need only 100 chill hours to set fruit. This makes them more tolerant of subtropical, warm and temperate climates than other deciduous fruit trees.*

# VARIETIES

Persimmons are small, easy-to-grow trees and can be categorised as non-astringent (sweet) or astringent (bitter).

The fruits of non-astringent varieties can be eaten both when they're firm and when they soften. This means that you can enjoy your harvest over a longer period. When harvested firm, the fruits can be stored for more than a month in the fridge. I personally prefer non-astringent persimmons and grow **Fuyu**, a semi-dwarf variety with a weeping habit. Other non-astringent varieties to check out include **Izu** and **Jiro**.

Astringent varieties, on the other hand, have fruits that are extremely bitter when firm and that cannot be eaten until they've fully ripened and are soft like jelly. Surprisingly, once the fruits are soft, they taste even sweeter than those of non-astringent varieties! Astringent persimmons are also more suited to cooler climates. But the biggest drawcard is that, because the fruits are so unpalatable while hanging on the tree, birds and wildlife will leave them alone. If astringent persimmons sound like your thing, then have a look at **Hachiya** or **Dai Dai Maru**.

Most persimmons are self-pollinating. However, having another persimmon nearby usually helps to increase fruit yields.

# GROWING

Grafted trees are available from your local nursery. The best time to buy and transplant them is during winter while they are dormant. When transplanting, be extra gentle because persimmons don't like their extremely sensitive roots to be disturbed. Stake your tree to encourage strong growth, and water immediately after planting.

Locate your persimmon tree somewhere with full sun, although it will tolerate a little shade. Make sure that the spot is protected from wind, as persimmon branches snap easily – especially when loaded with fruits – and fruit skins can become damaged if they rub against branches and leaves.

Persimmon trees require well-draining soil, as frequent waterlogging can cause their roots to rot. Plant your tree in a raised bed if you have a heavy clay soil. Young trees need to be watered regularly for the first two years until they have become established.

It's also important to water your persimmon tree during the growing season. Water consistently from early spring (to promote new growth and fruit set) to early autumn (to ensure large and yummy fruits). Water stress can lead to fruit skins splitting or fruits dropping.

**Flowers and fruits** Persimmons flower much later than most deciduous fruit trees. My persimmon tree is the last to flower out of all of the fruit trees in my garden, and at one point I thought it was dead! So don't be alarmed when all your other deciduous fruit trees have already bloomed. Your persimmon tree is probably just enjoying a long-deserved hibernation, which coincidentally also protects its flowers from any unexpected late frosts.

Grafted trees usually start fruiting after three years. It's best to remove any baby fruits before then to encourage the plant to focus on establishing its structure.

**Pruning** Persimmon fruits grow on the tips of the current season's growth. Because of this, you don't need a large tree to obtain lots of fruits. Big persimmon trees have more empty, wasted space inside the canopy. Persimmon branches are quite delicate, and long branches will snap if they can't support the weight of the fruits. Hence, keeping your tree to around 2 metres (7 feet) tall and wide is a good idea.

It's best to prune your persimmon tree every year, as unpruned trees can start to bear biennially (lots of fruits one year, and no fruits the next). Young trees can be pruned while dormant in winter, to help develop their framework. More mature trees can be pruned in summer.

When pruning, remove any dead, diseased or damaged wood. Remove any crisscrossing branches to prevent future fruits from rubbing against each other. Cut back extra-long branches so they don't later snap. Finally, cut back branches that have already fruited.

## HARVESTING

Persimmons are ready to harvest around autumn, when the fruits have turned their final, mature colour (deep orange) and are firm but slightly soft to the touch. Both non-astringent and astringent persimmons can be harvested at this stage and will continue to soften and sweeten after harvesting.

## STORING

Both non-astringent and astringent persimmons are best stored at room temperature, where they will naturally soften and ripen over time. However, if you want to slow down this process, then you can pop them in the fridge. Firm persimmons can be stored in the fridge for up to a month. Refrigerating ripe or soft persimmons will extend their life by only a couple of days.

*Persimmon fruits can be seedless, but if the persimmon tree is pollinated (even though it doesn't need to be), then the fruits will contain seeds. The first time I bit into a fruit with seeds, I thought they were bugs!*

# COOKING

Persimmons are delicious eaten fresh, so you can enjoy them just as they are. With non-astringent varieties, you can eat them at any stage from firm to soft like a ripe peach. I personally like them while they're still firm, but I also love the fact that you can enjoy them in a different way as the fruits start to soften. With astringent varieties, you'll need to wait until the fruits become as soft as an overripe tomato before eating. These persimmons are easiest to eat by cutting off the tops and scooping out the flesh with a spoon.

If you're lucky enough to have a glut of persimmons, then the best way to preserve them is by dehydrating the fruits. Here are a couple of ideas to try.

**Preserving soft persimmons: fruit leathers** Wash the persimmons (you can use either soft, ripe astringent persimmons, or soft non-astringent persimmons) and cut them into pieces, removing the stalk and any seeds. Blend until puréed. Spread onto baking paper or a non-stick dehydrator sheet, and dry at 50 degrees Celsius (120 degrees Fahrenheit) in your dehydrator for six to eight hours (or at the lowest setting in your oven) until dry but sticky like a roll-up. Cut into smaller pieces, roll up, and store in an airtight container in your pantry for up to three weeks or in the freezer for up to a year.

**Preserving firm persimmons: fruit chips** Slice the persimmons horizontally (firm astringent and firm non-astringent persimmons can be used) into pieces approximately 5 millimetres ($^1/_5$ inch) thick. A mandoline is a handy tool for this job. Dry at 50 degrees Celsius (120 degrees Fahrenheit) in your dehydrator for eight to ten hours (or at the lowest setting in your oven) until dry. Rotate the trays now and then to ensure even airflow. You'll know the chips are dry when they're crisp. Store in an airtight container in your pantry for up to six months.

When firm astringent persimmons are dehydrated like this, the process breaks down the tannins and the bitterness disappears like magic!

# YUZU

*Citrus junos*
Mandarin: 香橙 – xiāng chéng
Japanese citron • yujanamu

| | | |
|---|---|---|
| 🏷️ | **PLANT TYPE** | Evergreen perennial |
| 🌰 | **PLANT FAMILY** | Rue family (Rutaceae) |
| ⤡ | **PLANT SIZE** | Up to 4 m (13 ft) tall and 4 m (13 ft) wide but can be pruned to 2 × 2 m (7 × 7 ft) |
| 🪴 | **POT FRIENDLY?** | Yes, one tree in a 40–50 cm (16–20 in) pot |
| ✷ | **SUN** | Full sun |
| ❋ | **FROST** | Frost tolerant |
| 💧 | **WATER** | Regular watering |
| ❖ | **FOOD** | Heavy feeder |

Yuzu trees are beautifully evergreen and have unique leaves that look like one big leaf attached to one small leaf. They bear a citrus fruit that is popular in Japan, where the trees have been grown for thousands of years. It's an extremely seedy fruit that's not eaten fresh; rather, its juice and zest are used like those of a lemon or lime. Yuzu has a distinct, floral, grapefruit flavour with hints of mandarin and lemon, and it's a highly sought-after ingredient in the restaurant world.

I love yuzu with sashimi and fish, and I adore it in desserts and drinks. However, I haven't yet seen a fresh yuzu fruit in an Asian grocer, which is why I grow my own. They can be cultivated in the same way as other citrus plants but are more cold tolerant, making them a great choice for cooler areas. In Japan, yuzu grows on mountain slopes, and the cold weather improves the flavour.

*Yuzu trees have intense thorns that are much bigger than those of other citrus trees. Position your tree somewhere out of the way, so you don't accidentally poke yourself every time you walk past.*

# VARIETIES
In Japan, there are a number of yuzu varieties available that differ in size and flavour. However, outside of Japan, an unnamed yuzu will probably be your only choice.

# GROWING
Start with a grafted yuzu tree (dwarf rootstock is best for smaller gardens) from the nursery, as it will fruit within two to three years, and the yield will increase as the plant matures. Yuzu can also be grown from seed, but these trees will take up to ten years to fruit.

Like all citrus trees, yuzu will happily live in a pot, making it easy to fit into any urban backyard. I personally like growing my yuzu in a pot to contain its size and to limit the number of fruits it produces; I don't need many fruits for the amount of juice and zest I use in the kitchen. However, yuzu plants can also be grown in the ground and easily pruned back to 2 metres (7 feet) for easy picking. Make sure that you grow yuzu in full sun and sheltered from winds.

For a citrus, yuzu is extremely frost tolerant. The tree itself can withstand cold weather down to −9 degrees Celsius (16 degrees Fahrenheit), although the fruits are a little more sensitive. If you live in a climate with heavy frosts, then it's best to move the tree indoors during winter, and then pop it back outside in spring. Yuzu trees have the most fragrant blooms, which bees love. For everything else, follow the cumquat-growing guide (see pages 265–6).

# HARVESTING
Round and bumpy yuzu fruits are ready to harvest once they reach the size of a lemon. They can be harvested green, yellow or anywhere in between. Green yuzu is harvested in summer and has a tarter flavour. Yellow yuzu is harvested in autumn and has a sweeter, floral flavour. When harvesting, be mindful of the tree's large thorns – long sleeves and sturdy gloves can help you avoid being pricked!

# STORING
Freshly picked yuzu fruits can be stored in the fridge for up to three weeks, after which they will start to slowly lose their flavour. An easy way to preserve your harvest for longer is to freeze the juice in ice-cube trays. It will store in the freezer for up to six months.

To preserve yuzu zest, first zest fruits via your favourite zesting method, then freeze as a thin layer on an oven tray. Once frozen, break into portions and transfer to a re-usable freezer bag. The zest will store in the freezer for up to six months.

# COOKING

Yuzu is used to flavour dressings, sauces, drinks, vinegar, preserves, sorbets and more. It's also delicious in baked goods and desserts. It can replace lemons, limes or cumquats in any recipe.

## Homemade yuzu ponzu

Ponzu is a citrus-based Japanese sauce. It has a watery consistency and can be used in a multitude of ways: as a dipping sauce for dumplings or soba noodles, drizzled on fish and meat, or to marinate sashimi for poke bowls. This recipe makes enough for a couple of people, so scale it up for a larger dinner party. To store the ponzu, you'll need a clean and sterilised glass jar with a lid (see page 72).

Place the ingredients into the jar, place the lid on the jar, and shake well. Store the ponzu in the fridge for up to 1 week.

**MAKES 45 ML (1½ FL OZ)**

1 tablespoon Japanese soy sauce

2 teaspoons yuzu juice

2 teaspoons rice vinegar

1 teaspoon mirin

## Yuzu vinaigrette dressing

Yuzu is delicious in vinaigrette. I love to drizzle this dressing on freshly picked winter salad greens, and it also pairs beautifully with fish (try it on ginger soy barramundi with coriander; see page 219). This vinaigrette recipe makes enough for two large salads that serve four people each. To store the dressing, you'll need a clean and sterilised glass jar with a lid (see page 72).

Place the ingredients into the jar, place the lid on the jar, and shake well. Store the vinaigrette in the fridge for up to 1 week.

**MAKES APPROX. 100 ML (3½ FL OZ)**

1½ tablespoons olive oil

1 tablespoon rice vinegar

2–4 teaspoons yuzu juice

2 teaspoons sesame oil

2 teaspoons light soy sauce

2 teaspoons honey or maple syrup

1 teaspoon yuzu zest

Sea salt, to taste

Black pepper, to taste

*Mirin is a mildly sweet Japanese rice wine. It's commonly used in Japanese cooking to make marinades and sauces (such as teriyaki sauce), and to flavour stir-fries.*

# Where to buy Asian veggie seeds

## AUSTRALIA
- 4Seasons Seeds (4seasonsseeds.com.au)
- Boondie Seeds (boondieseeds.com.au)
- The Diggers Club (diggers.com.au)
- Eden Seeds (edenseeds.com.au)
- Garden with Connie (gardenwithconnie.com)
- Happy Valley Seeds (happyvalleyseeds.com.au)
- The Seed Collection (theseedcollection.com.au)
- Seeds of Plenty (seedsofplenty.com.au)

## UNITED KINGDOM
- Chiltern Seeds (chilternseeds.co.uk)
- Kings Seeds (kingsseeds.com)
- Mr Fothergills (mr-fothergills.co.uk)
- Real Seeds (realseeds.co.uk)
- Sea Spring Seeds (seaspringseeds.co.uk)
- Sow Seeds (sowseeds.co.uk)
- Tamar Organics (tamarorganics.co.uk)
- Vital Seeds (vitalseeds.co.uk)

## UNITED STATES
- Botanical Interests (botanicalinterests.com)
- Johnny's Selected Seeds (johnnyseeds.com)
- Kitazawa Seeds (kitazawaseed.com)
- Rare Seeds (rareseeds.com)
- TomorrowSeeds (tomorrowseeds.com)
- Truelove Seeds (trueloveseeds.com)

# References

## BOOKS
- *Asian Herbs and Vegetables: How to identify, grow and use them in Australia* by Penny Woodward (Hyland House, 2000)
- *Earth Restorer's Guide to Permaculture* by Rosemary Morrow (Melliodora Publishing, 2022)
- *Commonsense Citrus: A hands-on guide to propagating and planting* by Ian Tolley (Ihabi Publications, 2017)
- *Growing Chinese Vegetables in Your Own Backyard: Grow 40 vegetables and herbs in gardens and pots* by Geri Harrington (Storey Publishing, 2009)
- *Home-grown Mushrooms from Scratch: A practical guide to growing edible mushrooms outside and indoors* by Magdalena and Herbert Wurth (Filbert Press, 2017)
- *Homegrown Tea: An illustrated guide to planting, harvesting, and blending teas and tisanes* by Cassie Liversidge (Griffin, 2014)
- *Incredible Edibles: Grow something different in your fruit and veg plot* by Matthew Biggs (Dorling Kindersley, 2018)
- *Land of Plenty: Authentic Sichuan recipes personally gathered in the Chinese province of Sichuan* by Fuchsia Dunlop (W.W. Norton & Company, 2003)
- *Permaculture: Principles & pathways beyond sustainability* by David Holmgren (Melliodora Publishing, 2002)
- *The Art of Fermentation: An in-depth exploration of essential concepts and processes from around the world* by Sandor Ellix Katz (Chelsea Green Publishing, 2012)
- *The Plant Propagator's Bible: A step-by-step guide to propagating every plant in your garden* by Miranda Smith (Cool Springs Press, 2021)
- *The Seed Savers' Handbook* by Michael and Jude Fanton (The Seed Savers' Network, 1993)

## WEBSITES

- Bega Valley Seed Savers (seedsavers.scpa.org.au)
- Permaculture Australia (permacultureaustralia.org.au)
- Specialty Produce (specialtyproduce.com)
- The Seed Savers' Network (seedsavers.net)

# Notes

- **Page 11** – Celtuce contains double the amount … : researchgate.net/figure/Content-of-important-dietary-vitamins-for-lettuce-spinach-and-kale-expressed-as_tbl1_353062924
- **Page 11** – Water spinach is a great source of vitamin A … : weknowwatergardens.com.au/blogs/news/kangkong-the-unsung-superfood
- **Page 11** – Broad beans have high levels of protein … : ncbi.nlm.nih.gov/pmc/articles/PMC9025908
- **Page 11** – Goji berries are said to improve … : ncbi.nlm.nih.gov/pmc/articles/PMC8051317/
- **Page 15** – The principles of permaculture: permacultureprinciples.com/permaculture-principles
- **Page 29** – Members of the legume family are renowned nitrogen fixers … : floridamuseum.ufl.edu/science/plants-that-pull-nitrogen-from-thin-air-thrive-in-arid-environments
- **Page 47** – And please check your local regulations … : agriculture.gov.au/biosecurity-trade/pests-diseases-weeds/protect-animal-plant/dont-plant-it?fbclid=IwAR2n6w99vtMSnz3DE ubCypd0Pp6UkW1OgkayDVHaisQlWx8kvTsXvhA7NMQ
- **Page 52** – Cabbage moths and cabbage butterflies locate mustard family plants … : jerry-coleby-williams.net/2021/09/12/fly-my-pretties-natural-biosecurity-for-your-brassicas-stand-by-your-cabbages-and-give-nature-a-nudge
- **Page 67** – If sales went down in one city … : fuchsiadunlop.com/the-preserved-mustard-index/
- **Page 68** – *Brassica juncea* mustards are often included … : diggers.com.au/products/mustard-biofumigant#:~:text=Brassica%20juncea%2C%20Brassica%20napus%20(covers,Dig%20in% 20at%2021%20weeks
- **Page 77** – Bok choy or pak choy?: en.wikipedia.org/wiki/Bok_choy; dpi.nsw.gov.au/agriculture/horticulture/vegetables/commodity-growing-guides/asian-vegetables/a-f/buk-choy-embrassica-rapa-subsp.-chinensisem#:~:text=Varieties%20of%20Brassica%20rapa% 20subsp,leaves%20and%20flattened%20rosette%20shape
- **Page 124** – Komatsuna: researchgate.net/publication/337532473_Komatsuna_Japanese_ Mmustard_Spinach_and_Health_Continued_of_Japanese_Diet_For_Longevity
- **Page 124** – It's considered a super food in Japan … : edobeautylab.com/komatsuna-the-leafy-green-superfood-from-japan
- **Page 129** – Amaranth greens are highly nutritious … : pubmed.ncbi.nlm.nih.gov/20355024/
- **Page 144** – Eggplant fruits quickly oxidise … : ishs.org/ishs-article/1319_22
- **Page 175** – The leaves are especially prized … : sciencedaily.com/releases/2015/01/150114101642.htm
- **Page 213** – In addition, chrysanthemum is a Chinese medicine plant … : ncbi.nlm.nih.gov/pmc/articles/PMC7602131
- **Page 277** – A jujube tree's dream home is a temperate climate … : apps.cals.arizona.edu/arboretum/taxon.aspx?id=305#:~:text=Trees%20should%20be%20planted%2030,fresh% 20eating%20varies%20by%20cultivar

# Acknowledgements

The book would not have been possible without the support of some very special people.

To my garden-loving social-media community, thank you for giving me a platform on which to share my experience and passion online. This book would not be here without your ongoing support throughout the years.

To Tom, thank you for being my constant support and personal cheer squad, and for giving me the green light to paint my biggest ideas onto our entire backyard, without so much as blinking an eye. Thank you, with my whole heart, for always being there for me, and for giving me the confidence and space to chase my dreams.

To Rowena, for being there from the very beginning. For being such a multi-talented contributor and the best sister I could ever ask for.

To Mum and Dad, thank you for inspiring me to connect with and celebrate our cultural heritage through gardening.

To our happy pappies, Toro and Oakie, for being such playful, happy-go-lucky bundles of joy, and for being such great office buddies, even if you're both napping on the job most of the day.

To my fellow permies and all my friends, thank you for indulging passionately together in the wondrous delight of nature and edible gardens. Every human I have crossed paths with adds to and guides my gardening spirit.

To Jessica Tam, thank you for being my double-checker. To Olwyn, Shirley, Angelo and Aunty Margaret, thanks for letting me capture your gardens. To Maddie Callow, for painting magic into my hair for shoot days, and always.

To the incredible team at Murdoch Books. It has been a joy to work with you in bringing this book to life. I've never felt so 'on the same page'.

To my publisher, Alexandra Payne. Thank you for making me a published author. For believing in me, and for your continual support, guidance and enthusiasm through every step of it all.

To Loran McDougall and Kristy Allen for bringing this book to life through your expert eye and attention to detail, and for embedding my soul into every page.

Dannielle Viera, thank you for your patience with all of my changes, and for weaving your editor's brush through the text yet somehow leaving without a trace.

Michelle Mackintosh, thank you for translating my personality into warm, friendly pages and delightful illustrations.

Deborah Kaloper, thank you for making everything so effortless and pretty, and for always knowing exactly what works.

Alicia Taylor, thank you for your creative eye. It was a joy to shoot with you, as you made things so easy and captured everything so beautifully.

To you, dear reader, for picking up this book in the bookshop, library ... however our paths collided, thank you for supporting my first book. It means so much to me.

And lastly, to Mother Nature, for giving us humans the greatest gift of all. May I give back to you as much as you give to me.

# Index

Published in 2024 by Murdoch Books,
an imprint of Allen & Unwin

Murdoch Books Australia
Cammeraygal Country
83 Alexander Street
Crows Nest NSW 2065
Phone: +61 (0)2 8425 0100
murdochbooks.com.au
info@murdochbooks.com.au

Murdoch Books UK
Ormond House
26–27 Boswell Street
London WC1N 3JZ
Phone: +44 (0) 20 8785 5995
murdochbooks.co.uk
info@murdochbooks.co.uk

For corporate orders and custom publishing,
contact our business development team at
salesenquiries@murdochbooks.com.au

Publisher: Alexandra Payne
Editorial manager: Loran McDougall
Design manager: Kristy Allen
Designer and illustrator: Michelle Mackintosh
Editor: Dannielle Viera
Photographer: Alicia Taylor, Connie Cao
Stylist: Deborah Kaloper
Production director: Lou Playfair

*Murdoch Books acknowledges the Traditional
Owners of the Country on which we live and
work. We pay our respects to all Aboriginal and
Torres Strait Islander Elders, past and present.*

ISBN 978 1 76150 024 4

 A catalogue record for this
book is available from the
National Library of Australia

A catalogue record for this book is available
from the British Library

Colour reproduction by Splitting Image Colour
Studio Pty Ltd, Wantirna, Victoria
Printed by 1010 Printing International Limited,
China

**OVEN GUIDE:** You may find cooking times
vary depending on the oven you are using.
For fan-forced ovens, as a general rule, set
the oven temperature to 20°C (35°F)
lower than indicated in the recipe.

**TABLESPOON MEASURES:** We have used
20 ml (4 teaspoon) tablespoon measures.
If you are using a 15 ml (3 teaspoon) tablespoon
add an extra teaspoon of the ingredient for each
tablespoon specified.

10 9 8 7 6 5 4 3 2 1

 MIX
Paper | Supporting
responsible forestry
FSC® C016973